I AM
UNIVERSE

**The Universe Speaks With
One Voice To All,
Men Give It Many Names**

E. R. Margis

Published December 2010

INFINITY PUBLISHING
1094 New DeHaven Street, Suite 100
West Conshohocken, PA 19428-2713
Toll-free (877) BUY BOOK
Local Phone (610) 941-9999
Fax (610) 941-9959
Info@buybooksontheweb.com
www.buybooksontheweb.com

WARNING AND DISCLAIMER

On-going efforts were made during the writing and before publication of "I Am Universe" to make it's content accurate and free of errors. However there may be mistakes both typographical and in content. A new view of the universe is taken in this work, please be advised if you zip through the text by skimming you may miss some interesting content. The Universe, both within us, and outside of us is a great mystery, this book can be read like a mystery story, a piece at a time.

About this book's title

The title is not egotism, it is the universe speaking. A self aware universe, a psychologist might say it has "consciousness". As for the cosmological scene with stars, a comet and a group of asteroids, all of us tend to see what we wish to see, some see chunks of rock and a great void, others see god, and still others see themselves.

**Table of Contents
for the book " I Am Universe"**

Preface (preliminary remarks).......................... 6
Introduction 9

"The Cosmic Ocean", Three answers to the question, "Why did a Universe of Chaos,
Become orderly?".. 9

Questions from antiquity about our
natural world 13

Myths of Creation
        The Sumerian Creation Myth 20
        The Aryan-Hindu Creation Hymn........... 31

Supports or Worlds of the Universe............... 36
The Qualities (Gunas) of Things.................... 49
The Universe seen as "The Balance" 54

The Basic Components of the Universe.......... 66
The Tangible Universe................................... 68
Naming the Universe and It's 3 Components . 78

The Nature of Primordial Substance............... 84
The Nature of Energy 108
The Soul of the Universe............................... 116

Waves in the Cosmos 127
        The Electromagnetic Wave............... 132
        Standing Waves................................ 137
        Torsion Waves and this theory.......... 151
        The Nature of "Matter" 157
        The Artificial Electron 166
        The Gravitational Force 168

Traveling the Cosmos 180
The Phase of a Wave.............................. 189
Cycles in the Cosmos............................. 192

The Milky Way Galaxy............................ 193
        Orbits of the sun, earth,
        and earth's moon

The Two Selves of Man 230

Appendices

        Appendix **A** SI unit multipliers
                        prefixes and symbols 210

        Appendix **B** Poetic Wisdom from
                        the past 212

        Appendix **C** Five Complete Matrix Diagrams219
                        Physical Universe........... 220-221
                        Metaphysical Reality...... 222-223
                        Mythological World 224-225
                        "Mind" (Manas) 226-227
                        Psychological Self......... 228-229
                        Sumerian Mythology.......... 24-25
                        Aryan-Hindu Mythology......... 45

        Appendix **D** Conclusions on the nature
                        of waves 237

Glossary (Word Definitions) 239
                of Egyptian Goddess and Gods 260

Afterword Vishu dreams the universe 266-267

Index of persons mentioned in
                                "I Am Universe" 267

Index of subjects................................. 270

Preface
Preliminary Remarks

My hope in writing this book is to impart a greater understanding of the reality we find ourselves in. As with many things in life I believe our world, the universe, has a very simple foundation. Yet when it's three basic components interact, great diversity results, worlds of diversity.

These interactions take place in a cyclic manner. Cycles of nature, cycles of thought, cycles of creation. These cycles of change can be found recorded in every academic field, physics, metaphysics, psychology, and mythology. Because of these repeating cyclic patterns, understanding in two academic fields of endeavor can be compared and lead to a better understanding in both fields.

To facilitate this cross referencing of understanding in different academic fields, you will find what I term "matrix diagrams" in this book, The beginning of each diagram has a short family tree arrangement beginning with "the source" of all else in the diagram, this leads to the appearance of duality, then on to a group of three terms. This short family tree shows the origin of the three basic components that lead-up to the "matrix diagram" itself. Each matrix diagram consists of nine squares. The term in each square is related to terms in adjacent squares. The diagram squares look like those in the game "Tic-Tack-Toe". Matrix diagrams were also featured in my previous book "The Man Who Saw Space and Time" *

Matters not found in "I Am Universe" are:

Religion, Politics, War, Palmistry, Occultism, Astrology, Witchcraft, Messages from Spirits or Extra-Terrestrials, Human Reproduction Policy, or "Einstein theory" as **Poincare** ** termed it.

There is a minimum of mathematics in this book, mathematics is a tool used to describe activities in our world. Words can also describe our world, and when carefully used do so with scientific precision neatness and simplicity, in a word, with "elegance", also some situations or conditions are easier to grasp using word pictures rather than numbers. When wished words can also transmit warmth or coldness, a mood or feeling, grandeur or awe about a subject.

I believe the theory in this book is reasonable and is the result of applied common sense it presents a solid basis others can use to "push-off" from like a swimmer pushing off from the edge of a swimming pool. Due to the interlinked nature of the components and their characteristics, in explaining this subject some repetition occurs in the text. Please bear with me in this matter it was done with the best of intentions for clarity.

Considering this theory is in it's infancy, I invite those who believe they can improve it's usefulness to do so. If you are an amateur, that is a person who does not earn a living at the field you are interested in, do not be discouraged, remember science was begun by amateurs. We are all like men groping our way in a fog. This theory offers signposts along the way.

As a practical matter if you are an amateur, do not give-up your day job. Somewhere out in the world today, perhaps amongst one of my readers will be, a new Nikola Tesla, a new Sir Oliver Lodge, or a new Isaac Newton.

I wish you well in your search for a better understanding of our world and of ourselves.

I offer you a line from, Shakespeare.

"Angels and ministers of grace defend us."

This line is Hamlet speaking in
the play "Hamlet"
by William Shakespeare
Entered in Stationers' Register 1602

Sincerely,

E. R. Margis, September 2010

* "The Man Who Saw Space and Time"
by E. R. Margis, 2008, ISBN #07414-4386-4. published by Infinity Publishing. com.
It can be obtained by ordering it from your local book store or by calling the publisher at
1-877-289-2665. or from Amazon.com.

** **Poincare** – Jules Henri (1854-1912, French mathematician and philosopher of science). Poincare developed ideas on "relativity" independent of A. Einstein.

Introduction

Within us we have an inherent, a built-in, curiosity to learn about the world we find ourselves in, a curiosity born of a desire to survive, to "fit-in" better, to "get-along" with nature rather than fight it. We are strangers in a strange land and yet we are a part of it, searching to find our way from un-knowing to illumination, to making known the un-known.

Answers others have given to a few questions will let us see our world in the light they have, it is like viewing a scene from different angles, from each position the scene looks a bit different, some things become more evident whilst others become less evident. Answers to questions like: What is our world made of? What powers all motion? How long has the Universe existed? and more. Good answers are a guide from the oracle of the mind. a guide into the future,

Let us now go west of the moon into the past, and east of the sun, to days that are "coming-to-be" and view speculations about "The Cosmic Ocean".
Speculations on that vast and diverse reality like:

"Why did a Universe of Chaos, Become Orderly?

Over the millennia even with man's limited experience the following three answers have been offered, each from a different viewpoint.

Aryan-Hindu Mythology's answer

a) There is a tale in **Hindu mythology** that answers our question. In this tale of creation the god **Vishnu** is sleeping in the eternal Cosmic Ocean. Vishnu dreams the dream of universe. From Vishnu's navel grows the lotus of the universe with a beautiful blossom on which sits the creator god **Brahma**.** When Brahma opens his eyes a universe of order appears, when he closes his eyes the universe fades back from whence it came. When Brahma opens his eyes a universe of order appears, when he closes his eyes the universe fades back from whence it came.

Brahma lives for four hundred and thirty two thousand years, and when he dies a new Brahma takes his place, and when he opens his eyes a new universe of order appears.

Conclusion:

The Will of Brahma brings about order
in the Universe.

But there are two more answers to consider to the question "Why did a Universe of order appear out of Chaos?"

The Jewish Mystic's best answer:

b) This answer comes from a collection of oral traditions, opinions, and explanations and made it's appearance in the 1280's A.D. *** It is termed "The Kabala" (also spelled Cabala, or Qabbalah). The main work of this collection is the "Sefer ha-Zohar", "The Book of Splendor", among the lesser works included in this collection is "The Sifra di-zenti'uta", "The Book of Concealment". Amidst these writings, is the answer to our question. The reason a universe of order (Cosmos) emerged out of a universe of chaos.

The Answer is:

"God Wished to Behold God"

This is perhaps man's best effort to speculate as to why a god would make order out of chaos. I believe the answer could be restated as:

"God wished to experience God"

and from this logically I believe it could be said:

"God is the Universe and the Universe is God".

* **"The Logos"** Greek philosophers understood "The Logos" as the energy, the power to govern, to regulate, to bring order to the universe. I see these characteristics of the Logos as those of the Hindu gods Brahma and Shiva interacting through Vishnu with the substance of the cosmic ocean.

** **Brahma** - The god that brings order to that which is chaos (see the After word and illustration Fig. #46. pages 266-267. Brahma is one of the three gods of the Hindu triad (trimurti), the other two gods of the triad are Shiva and Vishnu. Brahma is not to be confused with, Brahman the non-dual eternal and absolute entity, of Hindu Vedanta, nor with the Hindu priestly caste who call themselves, the Brahmans.

*** 1280 A.D. falls in the "Middle Ages" of European history which extended from about 500 A.D. to 1500 A.D. and is also known as "Medieval Times".

Shiva - The god of dissolution. The destroyer that makes way for the new.

Vishnu - the god that sustains conditions in the universe. I see Vishnu as the balancer of the creative forces, and the forces of dissolution, The balancer of the wills of Brahma and Shiva.

Here is a third answer to the same question
"Why did order appear out of chaos?"

c) This is the answer a scientist might give it does not involve goddesses or god(s). To a scientist the volume of space that the universe exists in is vast and the duration of time of it's existence is probably unending, that is, it is eternal, and during that eternal existence the components of the universe have ample time to interact in all manner of combinations, proportions and configurations. So it is only a matter of time before the right combination of interactions takes place and order is born, an order that begets order. So to the scientist the answer is:

**"Order" (Cosmos) came to a chaotic universe
due to chance and probability.**

Which answer do you like best? The cold logic of the scientist, the mystical answer from Kabala, or the Hindu tale of the sleeping god's dream?

There follow questions about the universe written between 8 and 5 B.C. from the biblical books of Genesis, Ezekel, and Job. It is thought some were taken from older material and probably written by more than one author, each author adding what they believed were "improvements". There is also a statement similar to the creation account in Genesis I, taken from the New Testament book of John written about 120 A.D. it speaks of the Greek concept of **"The Logos"**.

Questions from Antiquity about our Natural World

The way a person states their question can be interesting for we ourselves may not have thought of it that way, we may think how wonderful it is to ask the question like that, or to think, if that is so, what then is the answer?

In the biblical book of Job Chapter 38 some marvelous questions are posed. The author(s) of Job are no longer known, but their questions live on. The book of Job was thought to have been written between the 7th and the 5th centuries B.C.

a) "Whereupon were the foundations of the earth fastened? Or who laid the cornerstone thereof; when the morning stars sang together, and all the sons of God shouted for joy?"
Job vs. 6-7

b) "Where is the way where light dwelleth, and where is the place of darkness:"
Job vs. 19

c) "Have you fitted a curb to the Pleiades, or loosened the bonds of Orion? Can you bring forth the Mazzaroth in their season, or guide Arcturus with his sons?" *
Job vs. 31-32

* The word **"Mazzaroth"** is generally translated as "Constellations". **Orion** (The Hunter) is a star constellation, the three stars that make-up the hunter's belt help one identify this constellation. The red super giant star Betelgeuse is at the hunter's right shoulder. The sixth brightest star in the sky is **Arcturus** in the constellation **Bootes** (The Herdsman). **The Pleiades** is an open star cluster four times wider than the full moon, six or seven of it's roughly twenty stars are visible to the naked eye. It's brightest star is blue-white Alcyone.

After these cosmological questions here are some that conservationists and the "green people" those who wish to preserve the natural world from destruction, should like:

d) "Do you hunt the prey for the lioness or appease the hunger of her cubs, while they crouch in their dens, or lie in wait in the thicket?"

Job vs. 39-40

e) "Who provides nourishment for the ravens when their young ones cry out to God, and they rove abroad without food?"

Job vs. 41

f) "Who puts wisdom in the heart, and gives the cock its understanding?"

Please do not misinterpret my inclusion of quotes from biblical writings in "I Am Universe". I am not being a "bible-thumper" here. If we put aside the faked histories and slanted pitches used to enforce some religious authority's need to speak them, or their wishful thinking, we will come across paragraphs that conjure up some truly creative images which bear on the topics of this book. I will site three such images for you, as you will see their selection is not to promote this or that religion. Let us begin at the beginning as it were:

a) From Genesis 1, Ch. 1, vs. 1-5. We have a wonderfully mystical attempt to explain how the universe came into being. Note that it was created out of a preexisting "something", a cosmic ocean of chaos. The creative act was to bring a measure of order out of chaos, or as the author of the book of Genesis put it:

14

"In the beginning God created heaven and earth. And the earth was void and empty, and darkness was upon the face of the deep: and the spirit of God moved over the waters. And God said: Be light made. And light was made, and God saw the light that it was good: and he divided the light from the darkness. And he called the light Day, and the darkness Night:"

b) This next quote is from the book of Ezekiel Chapter 37. Ezekiel finds he is on a wide relatively flat open area, a plain of great expanse. Ezekiel is asked to prophesy over some bones. You may recall hearing the first line as included in an old hymn:

"Dry bones hear the word of the lord! Thus says the Lord God to these bones: See I will bring spirit into you, that you may come to life."

vs. 5

Ezekiel goes on to say

"..and as I prophesied there was a noise, and behold a commotion: and the bones came together, each one to it's joint. And I saw, and behold the sinews, and flesh came upon them: and the skin was stretched out over them, but there was not spirit in them. And he said to me: Prophesy to the spirit, prophesy, Oh son of man, and say to the spirit: Thus saith the Lord God: Come spirit, from the four winds, and blow upon these slain, and let them live again. And I prophesied as he had commanded me: and the spirit came into them, and they lived: and they stood upon their feet, and exceeding great army."

This quote points out an ancient belief that "Substance" of itself does not "live" it must have "spirit", that is the power of life, energy infused into it to make it "to live".

Imagine Ezekiel standing on the open plain as he makes his pronouncements, gusts of wind begin to kick up dust clouds. You hear a noise, "the noise" of dry bones clattering about. You see flesh once again coming onto them, they are beginning to stand, they have actually "come alive", an army in full battle dress, swords gleaming. The dead coming to life, dust blowing about partially obscuring your view...then suddenly a figure steps out of a nearby dust cloud and seems to look you in the eye. It, it is one of them. This is a scene worthy of a Hammer horror film.

My third selection is from "The Gospel according to John" The opening paragraph is a prologue, an introductory overture to the rest of the gospel. This prologue is thought to once have been a separate work, existing before the gospel was written, perhaps a hymn and adapted for use in the gospel. The Gospel according to John was written in Greek Language, composed by at least three persons about 120 A.D. In this gospel is introduced an idea from ancient Greek philosophy, **"The Logos"**. In the prologue the Greek word "Logos" is used and the translator that made an English translation substituted the English word, "Word" for the Greek "Logos". Logos however is more than just speaking a word.

Heraclitus (580c -540 B.C. Greek philosopher) said of the logos, that, logos was shared by all. **The logos** is the "Will", the "Energy", the Power of the universe (some would say the "Will" of God). The Logos is the immanent cosmic governing, regulating, controlling principle in the universe. Logos is infused in all substance, all things. The Logos has also been defined as "Reason" perhaps it is more accurate to say it is the reasoning power of the mind. Remember wherever "The Word" appears in the prologue the original Greek said "The Logos". Here then is the compelling mystical prologue to "The Gospel according to John":

"In the beginning was the Word, and the Word was with God, and the word was God. The same was in the beginning with God. all things were made by him: and without him was made nothing that was made, In him was life, and the life was the light of men. And the light shineth in darkness and the darkness did not comprehend it."
John, Ch. 1, vs. 1-5

The "darkness" of this prologue is where the light is not, where the "Will" is not, where the message of the gospel is not.

The logos of the gospel is directed at "ordering", that is "regulating" the way a person thinks about the world and conducts their life as a result of this ordering. This prologue is like a restatement of the beginning of Genesis 1, Ch.1, vs. 1-5, only instead of applying "The Logos" to the ordering of the Cosmic Ocean, in the prologue of John's gospel it is applied to the ordering of men's lives. The prologue also personifies "The Word" as a person, personifies he who's pronouncements are like a light shinning in the darkness, or if memory serves me like "The light of the World". He who is his pronouncements.

What ever you believe about the message of the prologue, as a piece of literature I think it is a work of literary genius, whatever else the writer did in life he was in top form when he wrote this prologue.

It has been observed of the natural world that nothing seems to be permanent, even rocks wear down and crumble. Is there anything in the natural world that is "Permanent" that is "Eternal"?

W. S. Cather (Willa Sibert Cather, 1873-1947, U.S.A. Novelist) has said that there is a "universal human yearning for something permanent, enduring, without shadow of change."

For as long as man has existed, he has asked "What is all this I am encountering here?" and has been writing down the answers for roughly some five or six thousand years (the earliest known writing). So what does he think "all this is"?

In the pages to follow we shall see if the sages found such permanence and order. Let us look first at the Sumerian speculations about the universe, and then move on to the speculations of the Indo Aryan-Hindus.

Note: Since myths often have multiple deities I am adding this note in explanation of their existence. If the multiple **Goddesses and Gods** of mythology are looked upon as aspects or characteristics of a single all encompassing deity there is no conflict. An analogy would be the multiple parts of a machine, the parts have many purposes and forms that makeup the whole. The instance of white light is similar in that it is made up of all the colors. Take away the parts and you no longer have a whole.

Ancient Sumer

Ancient Sumer existed in what is now southern Iraq. Ruins of the Sumerian civilization are still there.

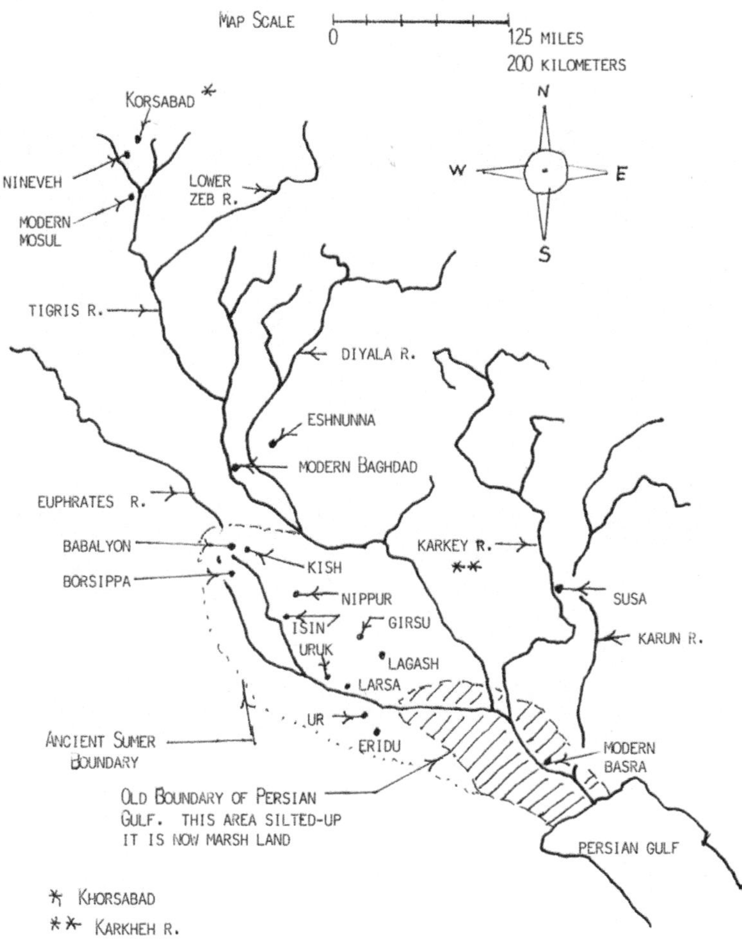

Fig. #1

The Sumerian Concept of the Universe

From "Apsu" the Cosmic Ocean came all. From Apsu came the Mother Sea "Nammu", Earth "Ki", Sky "An", and "Enlil" the power that directs order.

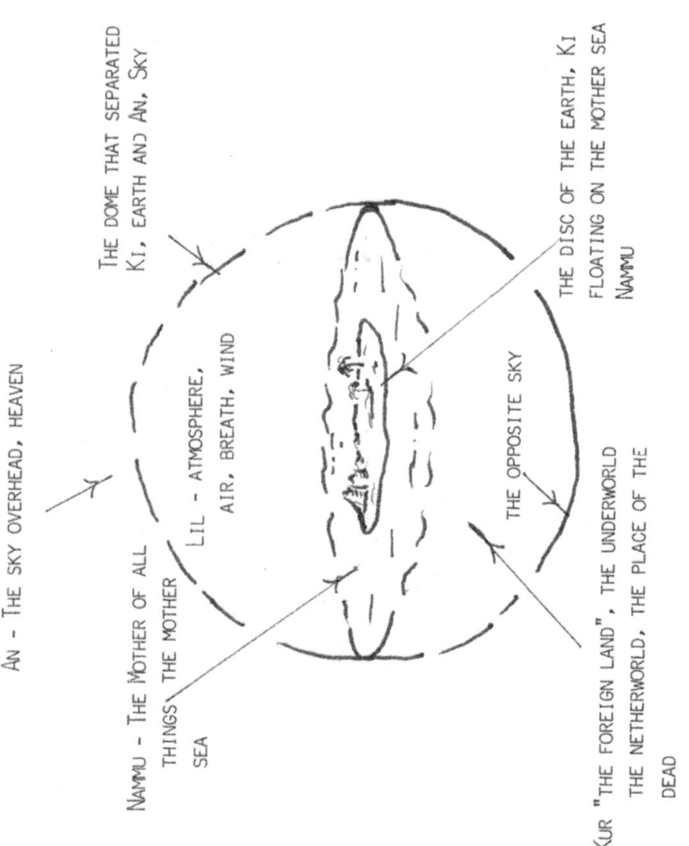

Fig, #2

The Sumerian Creation Myth

Sumerian sages believed before the earth was formed and the animals and man and the gods, a vast "Cosmic Ocean" existed that filled the universe. At that early time the "Cosmic Ocean" had no "things" in it, it is logical to assume from this that a state of chaos existed, a state of aimless disorder.

The Sumerian sages explained how the earth, the earthly seas, and sky (heaven) came to be, as the myth (story) goes, out of "**Apsu**" the vast Cosmic Ocean evolved "**Nammu**" the mother sea, the mother of all things, Nammu, the goddess of the abyss. From Nammu came "The Cosmic Mountain" called "Earth-Sky" **(Ki-An)**, earth and sky of the cosmic mountain were not yet separated, they were together.
A power emerged from Nammu that separated earth and sky. Sky (heaven) was thought to be masculine and earth feminine. The power that did the separating was personified as the god "**Enlil**". Enlil formed a union with earth **(Ki)** and from this union came all the other gods and goddesses and all created "things" of the universe.

Sumerian sages believed there were four cosmic elements, Earth, Water, Air, and the Sky a heaven of blazing stars.

The Creator gods of the Sumerians

These were "**An**" the mother sea, "**Ninhursag**" the power of regeneration, of new life, "**Enlil**" father of the gods, it was he that separated Earth and Sky and made possible further acts of creation, and **"Enki"**, the god that gave the power of reason and intelligence to men. As with many myths (stories), over the years they were "improved", that is altered to meet the needs of the myth writers. Improvements along the lines of, you want more prestige and money for the temple you are running give it's god or goddess a few more attributes etc. One goddess not mentioned in the creation myth was Inanna the goddess of love, love not for regeneration but for pleasure's sake, she was also the goddess of war. My own belief here is that "war" included not just physical combat but the ability to "Limit" who gets Innana's love.

Later Babylonians and Assyrians adapted parts of the Sumerian creation myth as did later Semites (Hebrews) as is evidenced in the Hebrew conception of the Universe, which is an adapted version of the Sumerians universe concept.

The Sumerian Vision Of Creation

One must say the Sumerian sages did a relatively good job of describing the creation of their world. They speculated as did the Aryan sages of India that a "Something" existed before creation of their world. The "something" of the Sumerian sages was The Cosmic Ocean **"Apsu"**. The Cosmic Ocean driven into liveliness by "Primordial Energy", driven into a state of chaos.

This relatively formless substance is termed "Primordial Substance" in the theory in this book.

The Sumerians believed that out of this formless substance all created things were formed, as do I. Out of this formless state eventually the mother sea **"Nammu"** appeared. Today one might say Nammu came into being by chance and probability of just the needed circumstances coming together in the Cosmic Ocean. Again out of chance and probability a Cosmic Mountain, a mixture of earth and sky as yet undefined, un-separated also appeared.*

After the appearance of the Cosmic Mountain out of "**Nammu**" the mother sea, came a directed power, a power capable of separating "**Ki**" the earth and "**An**" the sky, that power was personified in the god "**Enlil**". Once directed energy came into existence as "Enlil" his directed power was felt in "atmosphere" personified as "**Lil**", Air, Breath, Wind. Lil's breath is felt by the sea and birds and men. From the bright parts of atmosphere "Lil", emerged the sun, and moon, and the planets. Oh yes the Sumerians knew of the planets and their motions. They knew of six planets and earth's moon, plus one more a large planet "**Nibiru**", see footnote page 28.

Modern scientists and cosmologists do not improve all that much on the old Sumerian Creation myth, Oh! in place of gods and goddesses for directing and controlling the energies they have "Potential Energy" and "Kinetic Energy", they have "Velocity" and "Acceleration" as forces, and for the solid dome that contained the atmosphere modern scientists substituted the force of gravitation. Now it is known that the stars are distant suns, but science still wonders if the moon and planets affect human behavior (the pseudo science of Astrology claims they do).

* One can see the similarity here with the biblical Genesis account which speaks of the creation of heaven and earth (the cosmic mountain Ki-An), and then goes on to say, the waters were divided, those above were separated from those below. The realm of the stars above separated from the realm of earth below.

Matrix Diagram Goddesses and Gods of Sumer

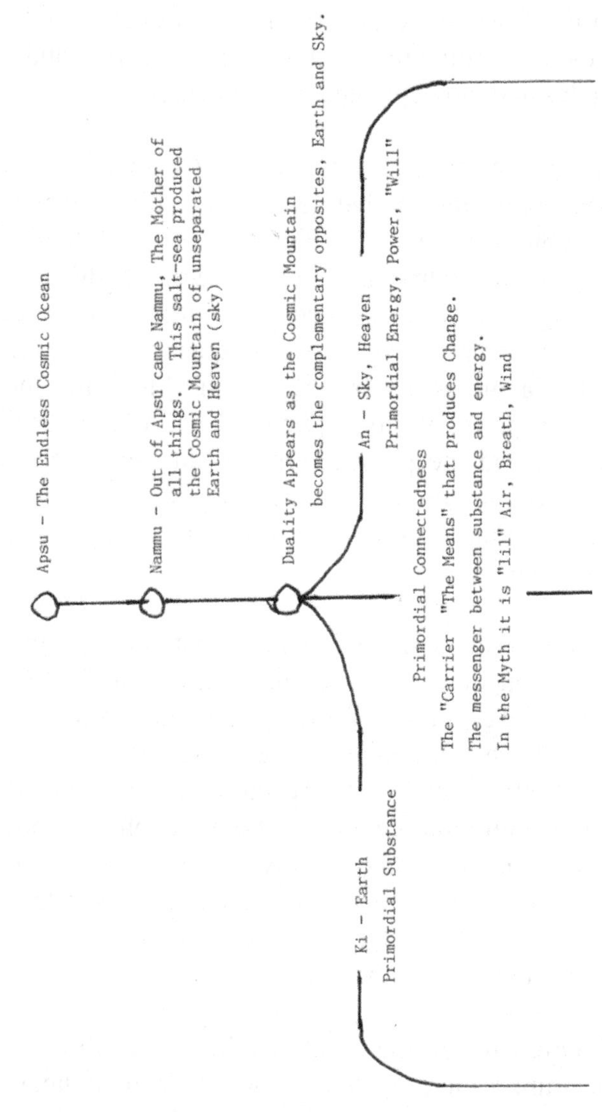

Fig. #3a

Ki -
Substance
The Dome of Heaven, The Sky Vault
Limitation, Permanence
Lil- Carrier of "the means" of change
An -
Power
Energy
Inanna Goddess of the morning and evening Star. Love as sexual pleasure. Love-War Strain in Space as the setter of form.
Ninlil - The avenging bride
Anunna (Anuna) - great gods
Ilu *
Definitions
Ninhursag (Nimmah) - Lady Birth
-"To Be" Whole
resultant condition (form or behavior of the moment)
The Complete cycle of change
Balance as alternaiton (sharing)
Sin - Nanna - moon god
Toarch of the Night
Time - At once young and old.
Enki - Change as force.
Anunnaki - great gods
Ilu *
Enlil - Utu - Sedu
The Sun
God of Justice establishes fates of Heaven and Earth
Decision, Directed Change
The Opposite Sky
The "Kur" (Hell), The Underworld
The Netherworld
Ershkagal - queen of the Underworld, Enki's twin sister
* ILu - an Akkadian word meaning a force of nature or a human power (force).
** Enki - The giver of shape to all things, the force that brins change.
*** Inanna - The desire for sexual pleasure, but not for purpose of procreation.
also - a goddess with the characteristic of being able to say "No" or "Yes". Limitations which either allow or deny.
The morning and evening star - The Planet Venus

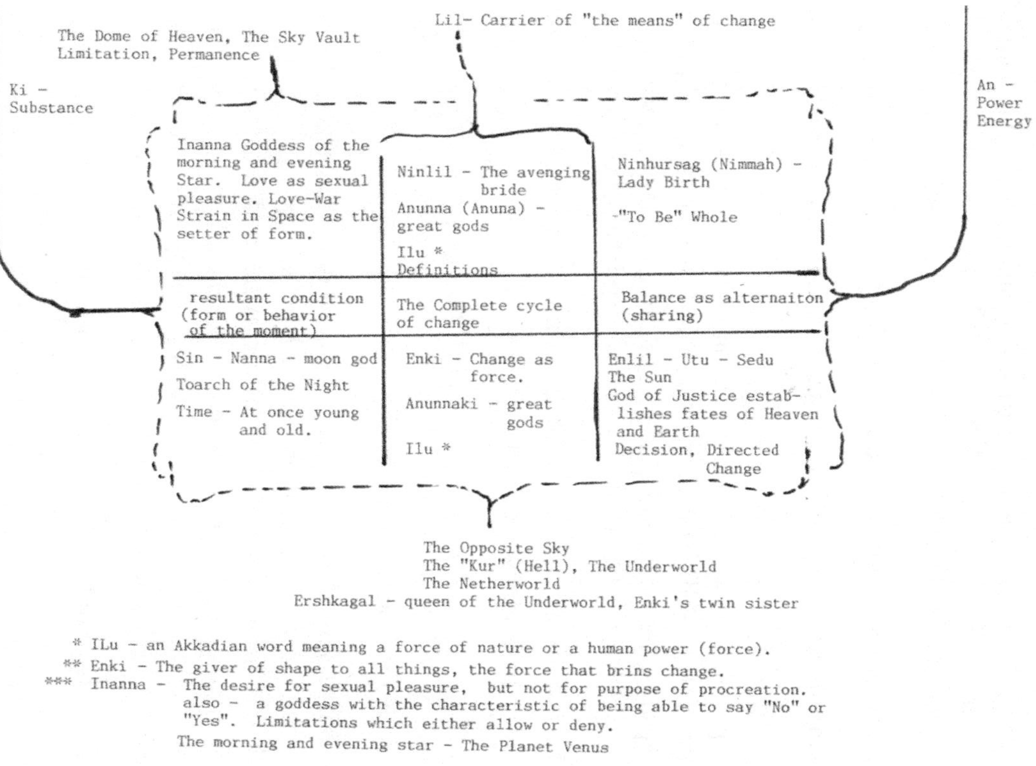

Fig. #3b

25

The Goddess and Gods of Sumerian
Myth and Cosmology

An - (Babylonian name: **Anu**) -, first among gods. **An** born of Anshar and Kishar. God who reigned over the heavens, who's name signifies "Sky".

Anshar – The celestial world.

Anunnaki – Divinities scattered over the earth and through the underworld.

Apsu -The Primeval Cosmic Ocean. In myth Apsu is the water that surrounded the earth and on which it floated. In this book Apsu is seen as the universe in a sate of undirected activity, over which the gods (the Divine powers) exerted their efforts of control.

Ea – born of Anshar and Kishar. A marine deity who's domain was Apsu. Ea means "House of Water".

Enki -lord of creation. As such in this theory Enki has been interpreted as the active "means" of delivering the force that "forms" substance.

En1i1 (Babylonian name, E11i1), Lord of the air. The god who separated sky and earth, from which he had been born. The god that is the **directed power** of air, wind, and breath. God of Storms. as such En1il the power to change what exists. The instigation of change, as an ability. (Babylonian names, Utu, Sedu) The power of the sun.

Ershkaga1 -Queen of the Underworld, the Netherworld. Sister of Inanna, In a myth Inanna asked Ershkaga1 to free her first love Dimmuz. Ershkagal refused and took Innana prisoner then later released her.

The Igigi – Divinities who peopled the sky.

Inanna (Babylonian name, **Istar**) Queen of the Evening and Morning Star (Venus). Said to be a goddess of love and war. I see Inanna as the pleasure aspect of sex. Sex for pleasure not procreation. She was free to encourage a suitor or rebuff an advance. Some of her suitors came to a bad end. In the theory in this book Inanna is also the limitation society places on a person and is one's own inhibitions as well.

Ilu -An Akkadian word meaning a force of nature or a divine being. Here in this theory Ilu is "the means" of implementing the force of change, it is unleashed directed energy. Ilu, the molder of Primordial Substance, and the force that molds the behavior of an individual.

Ki - Earth, (female) in the myth., Ki is taken in this theory to be "Primordial Substance", that which is formed by energy into all things.

Kishar – The terrestrial world.

Kur - The Underworld, The Netherworld, place of the dead.

Lil - Air, Breath, Wind

Nanna - (Babylonian name, Sin) -The Moon God. Here it represents the passing of Time, also opportunity. Chronos, Father Time.

Ninhursag (Babylonian name, Aruru) -Mother birth, motherly, nurturing. Ninhursag, Mother as the source of life. The need to be whole, brought about by male and female attracted to one another.

Ninlil - (Babylonian name, Mulitu) -The Avenging Bride. In the myth she was the bride of Enlil. Here she represents the recognition of limitation, the expectations of society and the individual. Enki did not honor those reasonable expectations.

Tiamat – A personification of the divine forces that bring about chaos, these are opposed by their complementary opposite divine forces that bring about order (Cosmos). In the myth it was from the union of Apsu and Tiamat living things were born. The Sumerians had other goddesses and gods I believe the ones given above are the most needed in their cosmology and myth.

Side Light: The Sumerians invented a system of writing using wedge shaped marks impressed in clay tablets (**cuneiform writing**) about 3,000 B.C. This writing system worked so well that others adopted it. The **Akkadians** (2350 -2300 B.C.) then the **Assyrians** (200- 612 B.C. and The **Egyptians** beginning about the time of Loser's step Pyramid 2680 B.C. to the time of Pepi I (2330 B.C.) onward to the time of, **Ramesses II** (1280 B.C.) It was this pharaoh that was the one thought to be in power at the time of the Jewish exodus. Thousands of these clay tablets have been found from these different civilizations. By the 1800's A.D. European linguists were able to read the cuneiform writing. The Egyptians also wrote on plastered temple walls, carved their writings in stone and used Papyrus paper to write on. Oddly the papyrus plant no longer grows in Egypt except in botanical gardens.

Footnote: The alleged planet **Nibiru** if it exists is speculated to travel in a very elliptical orbit moving in a direction around the sun opposite to that of earth (what astronomers term a retrograde orbit). Nibiru is thought to take some 3,600 years to make one complete orbit and travel at least six times farther out than Pluto in it's orbit.

The story told in the Babylonian text the "**Enuma elish**" tells of a collision (a battle) between Nibiru along with it's moons, and earth (called **Tiamat** before the collision). The result was a reduced size for earth and creation of it's present moon and the debris field known as the asteroid belt.

Today it is not known if the Enuma elish tale is just a myth, or fact, told in mythic fashion, in either case the tale is a whopper. Astronomers have however noted unexplained effects on the orbital motions of Uranus, Neptune, and Pluto that may well be due to an unseen massive body a great distance out from our sun. Nasa terms this alleged planet "**The Tenth Planet**" (**Planet X**). Author Zecharia Sitchin mentions this mystery planet in his books "The twelfth Planet" and "The Cosmic Code". The internet may have further information on the tenth planet.

About the Sumerian and Semitic Creation Stories.

In the Semitic account of creation of Genesis 1,Ch.1 we find that the light was separated from the darkness. This could be another way of indicating that chaos (darkness) was separated from order (light). The account goes on to say a "firmament" * was made to separate the waters above from the waters below. The word translated as firmament in one edition was in a later edition translated as "Dome".

In the Sumerian myth the water below this "Dome" or firmament was the water below the sky, Nammu the mother of all things. Nammu the sea on which the earth floated.

In the Sumerian myth the place above the dome, was high in the sky, the place rain came from.

Summer had some Semitic kings, such as **Sharrum-kin (Sargon) of Akkad**. An interesting legend exists about Sargon that he was an illegitimate child, his mother put him in a box on the river and it was found by the Gardner of the king of Kish, and by the favor of the goddess Ishtar he became vizier to the king. Does this tale sound familiar, could it be that it was reused by the Semites in their tale of Moses, who was put in a basket of reeds and set adrift in the river to be found and adopted by the royal family of Egypt. (Exodus ch.2 vs.3 and 10.)

* The word translated as **"firmament"** in the Duay Version. 1609 reproduced in P.J. Kenedy and Sons edition 1914, is translated as **"Domes"** in the Saint Joseph edition of "The New American Bible" © 1970. Genesis bk. 1, ch 1.

Sargon was in fact a former vizier to the king of Kish. Sargon dethroned king Ur-Zababa of Kish and defeated Lugalzaggesi of Uruk and became king of Sumer c 2350-2300 B.C.

So the Semitic sages came to know the Sumerian view of cosmology, therefore it is no coincidence that the Sumerian and Semitic Creation stories (myths) are similar. In this way myths of one people, are adapted by people of another culture, after all, a good story is a good story.

The Aryan-Hindu "Creation Hymn"

It is inaccurate to speak of a single, east Indian creation myth as the native people of India the Dravidians had theirs, and those that migrated to India the Aryans had their myths. It was in "The Vedas" the earliest written works in the Sanskrit language that the myths were put in writing between 1700 and 1100 B.C. and some of the Vedas were written as late as 6 to 4 B.C. Here we are interested in Rig-Veda mandala #10 which consists of 191 hymns including Nasadiya sukta #10.129 commonly termed: "The Creation Hymn".

This particular hymn comments on what existed before creation, and questions how creation came to be. I have only included phrases from this hymn that apply to the theory in this book. If you wish to view the whole of this creation hymn you might ask the Internet Search Engine to locate it for you. Even then, you may find the words in different translations vary a bit as each translator, choose the words they thought best represented what the original Sanskrit writer was trying to convey. The reader is left to choose the version they like best.

This particular Aryan-Hindu Creation Hymn speculates on the following:

1) What existed before "Order" took place, before creation.

2) How Creation came to be?

3) What first appeared of creation?

4) It questions if anyone will ever know when creation began, even the power that did the creating may not know.

Comments on The Aryan-Hindu
"Creation Hymn"

a) Before created things (order) there was "Neither non-existence nor existence then." I believe this speaks of a state of chaos in the primordial substance of the universe, the body of the universe, that which the Sumerians termed "The Cosmic Ocean".

b) "There was neither the realm of space nor the sky". In a state of chaos everything was mixed up, homogeneously disorderly and un-separated.

c) "There was neither death nor immortality then." If created things were not yet born, they could not die.

d) "Darkness was hidden by darkness in the beginning". If all was unpredictable, aimlessness throughout the entire universe it was like an impenetrable fog. Light here, being the informing of substance to be orderly. Darkness, being a lack of the knowledge of how to be orderly.

e) "Desire came upon the one in the beginning". Desire, being the motivation for change. "The One" being the power, the ability to cause change, the energy to cause change. The Power, that is "The One" already allowed chaos, desired order to come into being, and so separated the pairs of complementary opposites, such as Light from dark, order from disorder, substance from energy. The "impulse" the desire for this change, the motivation, is "Chance", and "Probability" amongst the chaotic interactions already taking place. Some interactions become repeated oftener than others, persist as order amidst chaos.

f) The writer of the "Creation Hymn" questions if the power itself could say when creation occurred.

If undisturbed, un-displaced time is eternity, it could be visualized as a circular band without beginning or end. A created thing appearing out of this endless eternity band would be like a disturbance on a quiet pond the only way the created thing knows where it is or how long it has been there is in relation to another disturbance that was on-going when it came into existence and still exists.

One cannot say when one's own creation began without being part of a previously created thing. My own conclusion to this reasoning is that "Creation" that is the ordering of the universe, is an on-going process. Disordering is also an on-going process. The two (creation and dissolution) being the positive and the negative sides of a cycle of change.

This "Creation Hymn" presents some wonderfully thoughtful statements, that beg a person to ask "What did the writer mean in saying that?" or "What needs to be the case, for this statement to be true?"

Aryan-Hindu Speculations on the Nature of the Universe

One off the grander Aryan-Hindu concepts is that of "The Planes of Existence" or "World planes".

There are six of these "Planes" or "Worlds", One is "The Unity", the complete sum of all the "Planes", It is termed **"Paran-atma"**, there is then a group of four planes of existence which I see as parts of a cycle of change, they are: **"Atma", "Buddhi", "kama", and "Sthula".**
There is one more "World" I have kept separate, it is termed by the Aryan-Hindus **"Manas"** (Mind).

Here in more detail are the Six Worlds or Planes of Existence.

Paran-atma also spelled **Param-atma**, The Supreme atma. It is the Supreme Existence, the totality of all that exists, has existed or will come to exist. In this theory it is the same as "Universe".
Paran-atma is also said to be the same as **"Brahman"**, the eternal reality, the Universe as the supreme self. In Vedic literature all is Brahman. Brahman has been described as "Absolute Consciousness". In the theory in this book it is Absolute Primordial Substance and Absolute Primordial Energy as well, it is the entirety of all that is. Paran-atma is a unity, imperishable and eternal, seen as a unity it can be said it is non-duel, yes true but one must ask, a unity of what? A unity can have components, after all, look at the human body it is a unity of existence, and can act in a unified manor yet it has parts, components. Some say Brahman is a state of pure transcendence, that which is beyond man's ability to experience Brahman. This is so as no man lives long enough,

or is vast enough or wise enough to experience the whole of Brahman, that is know Brahman over all space and all time.

An imperfect analogy would be a long running soap opera that a person had not seen the beginning or end of, only experiencing a brief bit in the middle, the person therefore would not know the whole story. *

* A "soap opera" is a serial tale told in one half hour or hour episodes a day at a time. some Have marvelous titles such as:

"The Young and the Restless", "The Bold and the Beautiful", "One Life to live", "Portia Faces Life", "Days of Our Lives", "The Right to Happiness", "As The World Turns", "Another World" (this one had 8,891 episodes), "Take the High Road", "Search For Tomorrow" (this one had 9,130 episodes), and "Gutezeiten, schlechte Aeiten" (Best of Times, Worst of Times). Wonderful titles all.

These programs have let people see how others have solved life's problems, some for the better and some for the worse.

The Four Supports or Pillars
Of The Universe

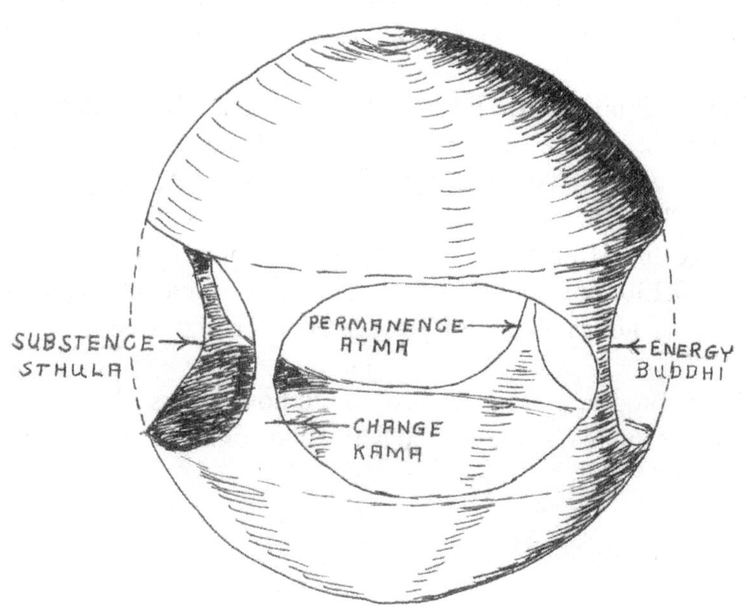

Fig. #4

The Supports or Worlds of the universe

The nine empty boxes in the center of this diagram represent the boxes of any "matrix diagram" in this theory. Each side of this diagram (consisting of three boxes) represents one of the four "worlds", Atma, Buddhi, Kama, and Sthula of the Aryan-Hindu World system. Above the diagram is the fifth world Paran-atma.

```
                    The Totality
                    Paran-atma    (Aryan-Hindu)
                    The Unity   (Ottoman Sufi)
                    Brahman    (Aryan-Hindu)
                    The Cosmic Ocean (Sumerian)
                    Ea        (Chaldea)
                    Ain Soph   (from Kabbalah)

             Atma  -  The World of Fixed Prototypes
                      The Celestial Realm
                      The Absolutely Invisible
                      The Plane of Nebulosity
                      The World of Defined Things
                      The World of Reaction to Change


Sthula                                           Buddhi
The Physical World                               The World of the Spirit
The World of Sensation                           The World of the Relatively
The World of the Now                                       Invisible
The World of Primordial-                         The world of Emination
   Substance                                     The World of Primordial-
                                                           Energy
                                                 The World of the Will



           Kama -  The Astral World
                   The World of Similitudes
                   The Angel World  (Angels understood as "swift messengers")
                   The Water Realm (place of the swamp of Desire)
                   The World that Imposes Forces
```

Fig. #5

Note: Corner squares, have characteristics that are shared by the two worlds adjacent to the corner square, Those shared characteristics are part of two worlds.

Four of the "Worlds" or "Planes" of existence I see as the four parts of a cycle of change, four stages of a cycle. These four are "Atma", "Buddhi", "Kama" and "Sthula". On the matrix diagrams of this book", each plane of existence occupies one side of the diagram (three square to a side). Adjacent "worlds" share the corner square terms.

<u>All matrix diagrams</u> in this book <u>share the same relative location of these worlds</u>. At the top of the nine square matrix is Atma, at the bottom is Kama, at the right side is Buddhi, at the left side Sthula.

See Fig. #17 page 77 for further comment on the sharing of the corner squares.

I refer to these four worlds as "supports" of the universe in the sense that these worlds support a cycle of change, support a process. They sustain the continuance of the activity of change, the cycle. Illustrated graphically they may be shown as pillars or columns as in Fig. #4 page 36.

Without change our world, our universe would have no motion in it, it would be a dead, frozen world. However our world is a changing, living reality, the entire universe is alive, it is a place where things come into being, and live out their lives, be they electrons, atoms, plants or men. The universe, itself in a manner of speaking "breaths". The universe over billions of millennia passes from a state of chaos (Disorder) to a state of cosmos (order), and with a mixture of the two extremes of existence between.

The old makes way for the new, the tired makes way for the young and the restless.

**The four "Worlds" of the cycle of
change will now be described.**

Atma –The permanent "Self" (thing). Atma is the divine world, the world where the Goddesses and Gods exist. Atma is the world of Permanence, The world of Ideal Forms, the fixed definitions of things and activities, Prototypes. Atma is the world of "Potentiality".

In the theory in this book "Atma" exists in all matrix diagrams even if not noted on the diagram. Atma is as it were, a step in a cycle of change.

In the psychology diagram (Diagram 5a-5b, pages 228-229), Atma is where Reason, Memory, and Imagination exist. Psychologically Atma is where we "Believe" what we believe. Atma is the world of the sub-conscious, the world of "The Shadow Self". "The Shadow" being the self which we believe we are not. Atma is the world of inhibitions, the things not to be gone beyond, not to be done, the inhibitions of society and of the individual. These "mores" help keep us from going, to far, to fast in our behavior. They are that we can compare "the new" to. As "Reason" Atma is the world where the mind makes sense of things and gives them value, for it is here that the complementary opposites, the "Ideal Forms" are matched into pairs, and in so doing all mixtures in-between these pure extremes, attain a relative value. Metaphysicaly, Atma is sometimes referred to as the world of the "Higher Self". The higher self being that part of a person's mind that is concerned with things that do not "run the body and maintain it", The Lower Self, does those things. The lower self is in the world of "Kama". Atma in this theory is where "Satisfaction" exists, the idea of heaven as reward is in Atma, existing as satisfaction, the desired ideal.

In the world of physics, Atma is where "Space" exists as "potential form", and inhibitions exist as "Acceleration", also Atma is the place of "Potential Energy". In Kabala, Atma is termed "The World of Creation". Creation in that, here diverse facts are made sense of, defined and given value as part of a pair of complementary opposites.

Buddhi -Buddhi means "To Awaken". Buddhi is the world where energy and power reside, where the "ability" to choose, to compare, to judge, resides. Buddhi is where energy is controlled and distributed. Buddhi is not life, it is the spark that causes otherwise dead substance "to awaken", to change, to act, to move. Buddhi is not breathing, it is the energy that powers breathing, it is the impetus "to breath" (The Connectedness carries the actual activity of breathing).

To a psychologist, Buddhi is "**Primordial Will**".
To a Mythologist Buddhi is "**Primordial Spirit**".
To a physicist, Buddhi is "**Primordial Energy**".

In the physical theory in this book, Buddhi is energy, Buddhi is attraction between complementary opposites, it is the repulsion of complementary likes, and it is a third thing, Buddhi is the need to share existence of opposites, a coexistence achieved by alternating domination of it's two complementary opposites characteristics so they co-exist harmoniously.

In realm of psychology, Buddhi is "The Need to be Whole" the uniting of complementary opposites like male and female, (this characteristic of Buddhi is located in the top square, right hand vertical column). Buddhi is also "The Need to be Individual", to be "self", to survive as self, to accomplish the needs of self, this characteristic of Buddhi is located in the bottom square, right hand vertical column.

Buddhi is also "The Need to Share Existence", to co-exist. This is accomplished by "Alternation" of, the need to exist of both complementary opposites in the Buddhi column. In human terms It is sharing through thick and thin, through joy and sorrow, pain and pleasure. All three of these psychological abilities, can be seen as "Love", love of simply "being" (as existence). Love as unity (two minds as one), and thirdly, love as mutual existence, a sharing, a coexistence).

In the matrix diagram of "Mind", Buddhi is "The Will". Buddhi as "Will" is "Reason" (top square of Buddhi column), and "Discrimination" (bottom square of the Buddhi column), Discrimination is the ability to separate, to choose, to make a decision and initiate that decision as a directed change. Buddhi is also "Wisdom" (center square of Buddhi column), Wisdom is "Harmony". Buddhi is both a "Potentiality" as "Reason" and an "Actuality" as "Discrimination"

Buddhi is a world you cannot quite put your finger on, it is a non-substance world. In Kabbalah it is termed "The World of Forming", forming in the sense of weighing and setting priorities, then initiating the unleashing power to achieve goals.

Kama – Is "Longing". Kama in Aryan-Hindu thought signifies "Desire", the word Kama means "Deed". Kama is the world of "Actuality" It is the world where the power of "The Will" exists as directed energy to satisfy physical needs, emotional desires, instincts, and passions. A world of continual on-going change, from Kama directed power is sent into substance, to form it, to mold it, to give it behavior. In the instance of psychology the forces of desire and need are sent to form the behavior of "The Self". Kabala terms this world "The world of action".

An Egyptian at the time of the pharaohs might have called "Kama", **"The Swamp of Desire"**, a place a person can sink into bottomless morass of desires and needs.

Medieval Europeans termed Kama **"The Astral World"** believing a person's "Fate" was regulated by the stars and signs in the heavens.

Emanuel Swedenborg (1688-1772, Swedish scientist, philosopher, theologian), termed Kama **"The Proprium"** that is the possession of, or attribute of a person.

The **Ottoman Sufis** believed God is one and is outwardly manifest and inwardly hidden. Ottoman Sufis termed Kama **"The World of Similitudes"**. A "Similitude" is a resemblance, or likeness, it can be seen as a double. My interpretation of this is that there is <u>a visible self, the physical self</u>, and there is <u>an invisible self, the self of our actions and choices</u>, our behavioral self only apparent over time, **The Double** to the physical self.

Others have termed Kama **"The Angel World"**, An angel by definition is a "Swift Messenger", and that is just what the on-going forces in Kama are, Swift Messengers. The messages carried are the desires, and needs, delivered by "The Connectedness". If you doubt the swiftness of delivery of these messages consider the phrase, "It was Love at first sight". Is that swift enough for you?

Kama in this theory is where the message of power is delivered into primordial substance to mold it, to form it. In substance that power becomes **"Stress Forces"** that form substance.

Sthula -Sthula means "Gross", gross in the sense of having bulk, volume. It is our everyday physical world, in which we struggle to survive. A Japanese might rightly term "Sthula", **"The Floating World"**, in that the forms in it are always changing, flowing from one shape, or state to another, carried by the river of time.

Sthula – Body. Sthula is where our accomplishments and errors become physical reality, where the potential to take form becomes an actual form which changes from one moment to the next.

In the Matrix Diagrams of this book "Sthula" is located in the left hand vertical column's three squares. The top square is the potential form of substance. Space possesses potential form. Space can be "molded" by the forces of energy, into any form. The bottom square of this column has the actuality of "Time". The Substance of Time which when displaced yields "duration" this gives all created "things" a life-span. The central square of this column contains the result of the interaction of displaced Space and displaced Time <u>revealed a moment at a time</u>, this is the moment of "the now", the moment of existence. Sthula is like a series of motion picture film still photos which together form a movie of life. The still frames flow one into the other to reveal the action. It is in this moment of "the now" that things exist.

In the world of physics, Sthula Is the world of physical reality, of sensation, of events, of experience.

In the world of psychology Sthula is the world of "Behavior". In mythology Sthula is where. man is subject to the whims of the god of opportunity, the whims of "Time", and the whims of the goddess that limits form "Space", who gives the limits a place "to be".

In the physical world Sthula is where the effects of the forces of **"Stress"** and **"Strain"** are resolved. **Stress**, the result of on-going directed force in the act of molding Substance, displacing substance, and **Strain** the result of substance having been molded, having been displaced.

In Kabala, Sthula is termed "The World of Making", a world where substance is formed. Sthula is where a thing has a life-span, it is where, young men become old men, where a world that is, becomes a world that was. Sthula completes the cycle of change.

Sthula is the experienced world. In the mind Sthula is perceived reality, it is memory, sensations. Sthula is in it's way, the gateway to Atma the divine world, the world of the Gods where new visions of what a person may or may not desire or need are pictured in the mind.

Shown in Fig. #5, page 37 are the correspondences between the four "Worlds" or "Planes" of the Aryan-Hindu system and those of the Hebrew mystical writings of Kabala, along with the author's interpretation of them.

In Fig. #6, page 45, can be seen the location of the five worlds that are in every matrix diagram in this theory. Each "World" occupies one side (three squares to a side) of the matrix diagram. The fifth world is the origin term at the top of the family tree.

The sixth world of the Aryan-Hindus is **"Manas"** (**Mind**). The diagram of "Mind" is shown near the end of this book on pages 226-227, Diagram 2a-2b. I have placed Manas, Mind in it's own matrix diagram because the thinking process is a cyclic process. The mind is a logic machine driven by perceived desire and need. Mind is the mechanism that regulates behavior, within the mind diagram you will see the four worlds of the cycle of change.

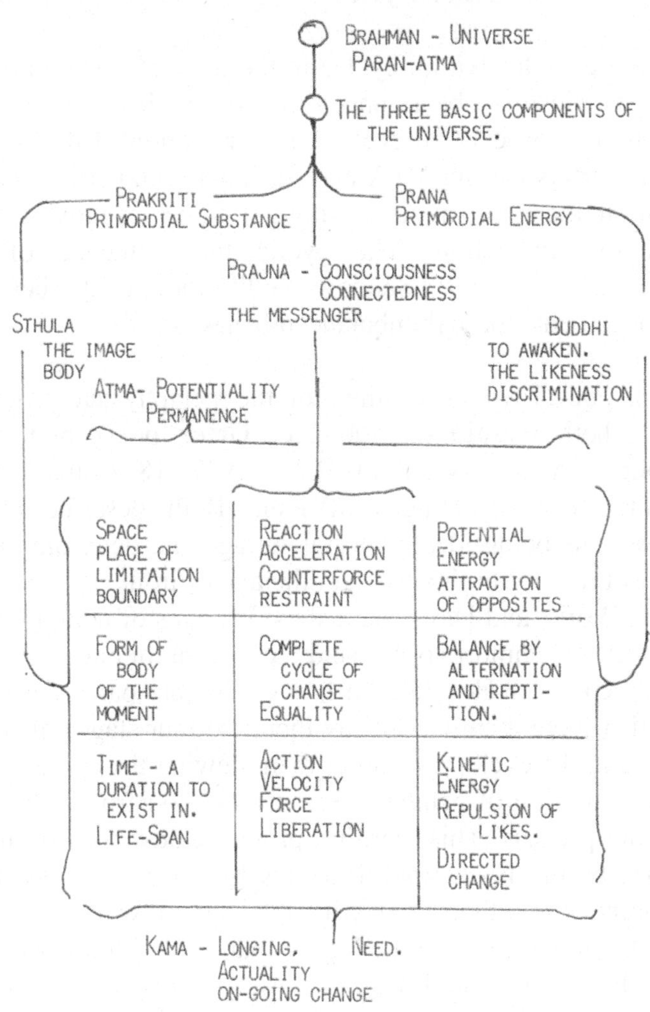

Fig. #6

Atma – The permanent self (thing), **Buddhi** – To awaken, **Kama** – Longing, **Sthula** – Body, these are the four parts of the cycle of change. In the squares are the characteristics of the three basic components.

Other Cycles

There are cycles within cycles in the universe. The universe itself is sloshing back and forth from chaos to cosmos (disorder to order), the cyclic circular motion of the planets in their orbits moving back and forth between perihelion and aphelion, the cyclic nature of thought processes moving from need to satisfaction. The cyclic the influence of the actualized Self and "Shadow Self", they cycle between balanced behavior and unbalanced behavior.

At this point I would comment on the metaphysical "Ages of Man", both **Hesoid** (c. 800 B.C. Greek poet), and **Ovid** (Publius Ovidius Naso 43 B.C. –A.D. 18 Roman poet) commented on the ages of man. Both describe man's behavior as being ideal in a golden age and deteriorating to animal brutality in the last age, the age of iron. Hesoid in his work "Works and Days" describes five ages of man, Ovid in his work "Metamorphoses" said there were four ages of man. Let us consider Hesoid's fifth stage as just part of a basic four. I believe then we can say there are four stages of social change in the cycle, and rather than viewing the cycle as one of deterioration of behavior, see it more constructively, as a learning process. This learning progress, advances from the simplicity of the animal learning to survive, to learning harmony with others and with the natural world. Cycling from the lower man, to the higher man. As with other cycles both the higher and lower are always present it is just a matter of, the proportion of each that is active, death) aspiring to be in balance with the ideal higher self of the metaphysical "**Individuality**" that survives death. I see this

Metaphysically as concerns "**The Self**" this cyclic activity can be seen as "**The lower Self**", the actualized self, trying to achieve the ideals of the "**Higher Self**". The lower self of this cycle is the metaphysical "**Personality**" (which perishes at

latter, as the effect the person's activities and works had on others and society. This cycle of social and personal development can be put in place in a matrix diagram of the Aryan-Hindu "World planes" as follows:

l) Kama - **The age of Iron**. The care and use of the physical self including it's survival.

2) Atma – **The Golden Age**. The ideal existence viewed as potential reality, as a thing that can come to be, as a goal in life.

3) Buddhi – **The Bronze Age**. This age is where intellect and reason sets itself to accomplishing the envisioned goal.

4) Sthula – **The Silver Age**. Accomplishment of the goal, however imperfect it may be. This age psychologically is ideally the existence of harmony between "The Self" and that which is not self, which is "Other" (as is the shadow self). Physically and ideally it is being in harmony with the natural world. The Silver Age is the result of trying to make "the potential ideal", actualized reality.

Sthula, The Silver age should not be looked down on, it is actualized reality, without it we would not exist as living beings. It is fine to have ideals as potentialities, but as a practical matter we must live as best we can.

We live life a moment at a time, we are never "complete" in that moment. We constantly seek to satisfy "Needs" and "Desires" that is part of being alive. We may envision ideals in life to be striven for but the actual goal is <u>to live in harmony with the world</u> and <u>what we can practically achieve</u> of the ideals.

To use an analogy, the real world is our "**bird in the hand**", our personal ideals are potentiality, they are our "**bird in the bush**". A bird in the hand as the saying goes, is worth two in the bush.

Each cycle of change has the same four parts.
These are:

    1) The ideal potential.   Atma.

    2) The power to bring change about. Buddhi.

    3) The on-going directed effort to
                    achieve the change. Kama.

    4) Accomplishment of the change as
      an actualized reality, a moment at
                     a time. Sthula.

The above four phases of the cycle of change support it, that is, carry it forward in time.

"The Qualities" or Gunas

Gunas are the fundamental qualities of all objects in the manifest world. These "Qualities" are three in number, two are complementary opposites, and the third is a way for the two opposites to exist in the world together, to co-exist, to as it were, exist in harmony.

The three Gunas are:

Tamas is **Potentiality** and a reaction to change.
Rajas is Actuality the instigation of change,
Sattva, is Harmony, a mutually beneficial sharing of existence, it is wisdom.

Tamas -a reluctance to change, to inhibit change. Tamas is the quality of heaviness lack of initiative, of procrastination (putting things off). In the world of psychology Tamas quality is inhibitions, inhibitions of the individual and inhibitions of society (society's mores). In the world of thought, Tamas is the fixed prototypes, the definitions of things. Substituting a word definition for the actual thing, usually takes less effort (a kind of laziness). In the theory in this book Tamas quality in the world of physics is possessed by "acceleration",

Rajas -Rajas is the initiating and sustaining of activity. Rajas, is the quality of lightness. In the world of psychology it is restlessness, is desire. Rajas is activity, action, aggressiveness. Activity to satisfy needs, emotions, and achieve satisfaction. It has been said of the Rajas quality that it is good or bad according to the amount of consideration given to others. Rajas is the quality of force that overcomes Tamas, overcomes the obstacle to realization of one's goal. In the world of physics "velocity" has the Rajas quality.

Sattva -Sattva is harmony. In the world of psychology, Sattva is the process of "Individuation", the balancing of an individual's personality between what they believe they are and what they believe they are not (their shadow self). Sattva in the world of physics is achieved by alternation and repetition, it is oscillation and resonance, in this way each complementary opposite can exist in harmony, in co-operation with the other, each has it's time to fulfill what the other is not.

Fig. #7 shows the Location on any matrix diagram of the three "Gunas" or "Qualities" drawn using the methods in this book.

Tamas	Tamas	Tamas
Sattva	Sattva	Sattva
Rajas	Rajas	Rajas

Fig. #7

Any term that occupies a square in the diagram will have the "Quality" of that square. <u>Each "Guna" or "Quality" occupies a horizontal row of three squares.</u>

The terms of the central horizontal row, are all resultants of the interaction of the terms in the row above and the terms in the row below the central row.

In the field of mythology, Sattva is a balance of the genders of the goddesses and gods, similar to the situation with the two selves of psychology only in mythology the gods do their own harmonizing for coexistence. In the Mind and the world of thought Sattva when achieved, is termed "Wisdom".

Side Lite: in the construction trades there is a device that exhibits the quality of "Sattva" it is **the "Water Level"**.

A water level is essentially a garden hose that is long enough to reach between the two points to be leveled, plus two or three meters additional length. The hose is filled with water, care being taken to not have any air bubbles in the water filled portion. Plug both ends, (a thumb over them usually works well). Two persons are needed, one at each end. Stretch the hose between the two points to be leveled, up-lift each end a suitable amount and note the level of water in the hose. Mark the water level on a wall, or with a stake in the ground, or simply measure it off a fixed point and jot it down. Use of <u>a transparent hose</u> near each up-lifted end, makes seeing the water level easier. This method will work inside or outside of a building, down a stair well and up the other side and also work around corners, A direct line of site between the points to be marked as level is not needed. The method is as accurate as a laser level, or a transit, but a little wetter to use. When done using a water level, the garden can be watered with this device.

Graph of a Simple Harmonic Wave Components.
and their Gunas

This theory is a wave theory, and as such the "Gunas" can be applied to a simple harmonic wave, a wave of the displacement of "Primordial Substance". In such a wave **"Acceleration" is a Tamas quality**, and **Velocity" is a Rajas quality**. **Sattva** is the quality of resonance in a wave' the sharing of existence by alternation.

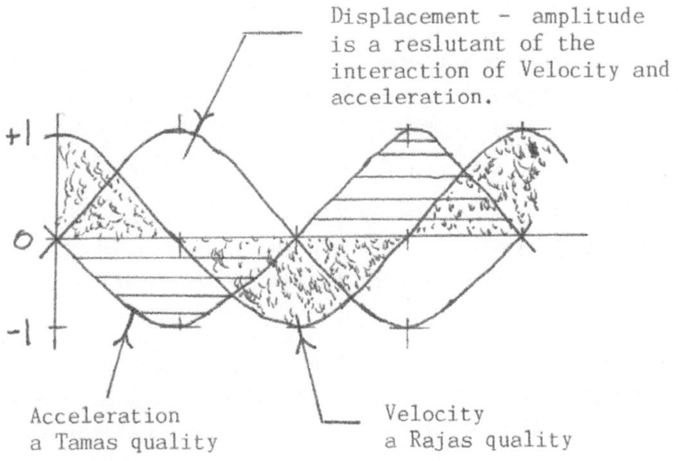

Fig. #8

Note Velocity and Acceleration maximums and minimums are never equal in the same moment. When one is maximum the other is Zero and Contra-wise.

Where Dynamic Equilibrium Lies
A place to park a spaceship or satellite

Lagrange (Joseph Louis Lagrange, 1736-1813, French-Italian mathematician), among his mathematical contributions to celestial mechanics, he noted that when one body orbits another there are five points in space where a third body can position itself in "**dynamic equilibrium**" (astronomers term them "Langrangian libration points"). At these five positions if a third body moves in synchronism with the orbiting body, it will "float along" as it were, in orbit around the sun, not gravitating toward either the sun or the earth. Each planet in a solar system has it's own set of Lagrange equilibrium points

Fiction writers placed A Twin planet to Earth at L-3, satellites have shown it is not there.

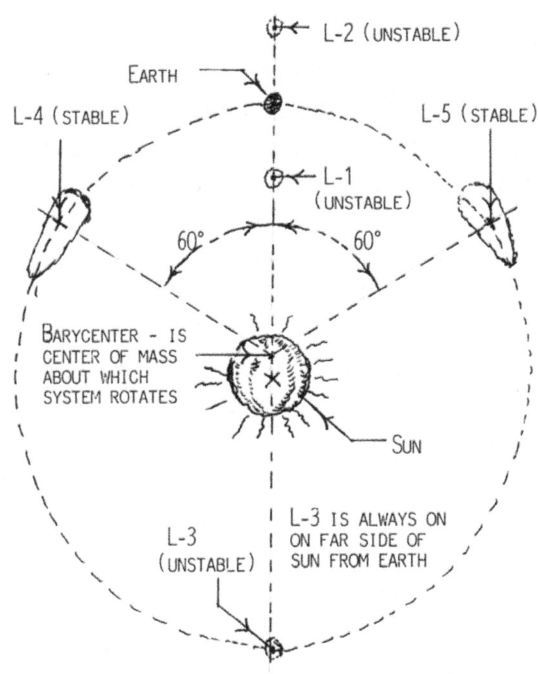

The Universe as "The Balance"

To Possess "Balance" two or more things must be equal in some common way. This equality may be one of force, weight, power, opinion, geometrical shape, or they may have a common equality in the way they differ.

There are two general types of balance, One is "**Static Balance**". Static Balance is where things even under changing conditions have a given equality that does not change.

The other type balance is "**Dynamic Balance**". Dynamic Balance is some sort of change balanced over a period of time.

The pairs of complementary opposites in the universe are by their nature, in "**Static Balance**". The two terms of such a pair are the extremes of their type both are what the other "**is not**". No matter what the magnitudes of either partner of the pair, their "Static Balance" remains unchanged.

Dynamic Balance in the Universe takes place over a period of time, an analogy that shows the character of dynamic balance is that of two trains traveling on parallel tracks the same distance to the same destination. One travels slow at first, finishing the trip traveling fast, the other travels fast at first finishing the trip going slow, each takes the same average time for the trip. They are time wise, in "Dynamic Balance". An automobile wheel and tire may be in perfect "static balance" when stationary, yet when run up to speed, be found not in "dynamic balance" because the weight distribution is not symmetrically placed on opposite sides of the center of rotation of the tire and wheel.*

To dynamically balance an out of balance tire and wheel, the mechanic adds a small weight as needed (clamped on the wheel rim) to achieve dynamic balance,

Are Simple Harmonic Waves Statically Balanced and Dynamically Balanced?

If a Simple Harmonic Wave only had "Static Balance" it would just "sit there" it would not "wave" at all. However we know it does "wave" and it's ability to do so must be in it's "Dynamic Balance" characteristic. The Energy in the wave keeps sloshing back and forth, from being mostly Potential energy which is energy stored in the "Strain Forces" due to displacement of the wave's substance, to Kinetic Energy, due to the motion of the wave's substance, back and forth, back and forth, over and over. The reason this takes place is The Displacement and Velocity amplitude maximums and minimums are 1/4th of a wavelength out of phase with one another. See Figure #8 page 52, graph of wave characteristics.

An analogy of this situation, though not perfect, would be of a coil spring when the spring is being compressed it's kinetic energy of motion is being transferred into potential energy of elastic strain forces in the substance of the spring. The spring then would be able to rebound (due to the elastic forces in it).

* for those not conversant with the terms of automobile parts, the "wheel" here is the central steel (or aluminum) part the rubber tire is mounted on.

As soon as the stress forces compressing the spring are reduced or removed the spring rebounds, in so doing the potential energy of strain stored in substance is converted back into kinetic energy of motion . In short, **A Simple Harmonic Wave is Dynamically Unbalanced during each moment of it's existence**, but over a complete cycle of change, it is "Dynamically Balanced".

All wave motion in **the universe is driven into activity and maintained by the existence of imbalance** <u>at any one moment</u>. From one moment to the next Waves are in "Dynamic Unbalance". Yet overall, that is over the period of time of each single the wave cycle (wave length) is in "Dynamic Balance".

The three characteristics of a Simple Harmonic Wave change amplitudes like the movements of dancers in a ballet, they complement one another. They have the overall quality of "Sattva" "Harmony". Each complete cycle of activity brings an **"Equality of Existence"** to it's characteristics and it's components

Examples of Pairs of Complementary Opposites that are in "Static Balance" by definition are:
Substance and Non-substance, Chaos and Order, Female and Male, Permanence and Change, Repulsion and Attraction, One and Many, Space and Time, Potential, Energy and Kinetic Energy, Same and Other, The Sub-Conscious
The Universe itself over a period of time s and the Id.

Examples of "Dynamic Balance" are:achieves dynamic balance between it's states of Chaos and Cosmos (Disorder and Order). An equality in the "presence" of it's two states. A simple harmonic wave also has this sort of dynamic balance.

The "The Balance" spoken of in Kabala, is a balance between Revelation and concealment, a balance between what is said openly to all and what needs to be interpreted by a teacher, who reveals the concealed meaning.

**Two speculations on the use of
controlled imbalance.**

1) The electron is a spherical standing wave structure in continual resonance. One of the latest speculations on the nature of matter is that protons and neutrons are both made-up of electrons. If the electron resonating wave imbalance could be modified it's non-resonant energy could be "siphoned-off" and put to use heating homes, powering mobile vehicles, or provide energy for industrial processes.

2) This second speculation has to do with **radio-active substances**, all the elements above #83 Bismuth are radio-active, and so is element #43 Technetium (formerly called Masurium, 1925 discovery claimed by German chemists I. Tacke, W. Noddack, and O. Berg). These naturally radioactive substances can be hazardous along with the fact that the Atomic-Electricity generating plants and all the facilities that process radioactive materials have huge amounts of radio-actively contaminated waste products. Would it not be nice if a method could be had to make these radio-active chemical materials un-radioactive when needed? In short, a method to make dangerous radioactive waste products safe, To **"de-tune"** the resonance that makes them radioactive and end up with more-or-less harmless by products.

This second speculation would have great value in the world today as not one ounce of radioactive waste has ever been permanently disposed of since the first atomic energy experiments and reactors. Examples of temporary disposal used today are:

1) Dumping in the ocean. I cannot see that this is doing the ocean any good, it may be doing the bottom line (finances) of the reactor operators a lot of good at the expense of the environment.

2) The radioactive remains of the first U.S. nuclear reactor that was operated in Chicago Illinois, are now buried under four feet of earth in a public park, and water wells in the park are now showing increased levels of radioactive tritium in park well water.

3) Stored in water pools at nuclear reactor facilities. They glow with a lovely blue colored light, (from Cerenkov radiation) ** and also continually heat the pool. These very limited capacity temporary storage pools are used to hold spent but still radioactive fuel rods.

4) Buried temporarily in trenches at nuclear refining and experimental facilities. There is some risk of rain water concentrating the radioactivity in the bottom of these trenches.

5) Disposal by direct Pumping of millions of gallons of radioactive water down unused water wells where it would disperse in with the natural ground water. The operators of a uranium refining plant asked their lawyers if they could dispose of millions of gallons of radioactively contaminated water down three unused water wells on the property and **their lawyers said "yes"** the reasoning being no one could positively trace where the radioactivity had come from once it was mixed in with the natural ground water. Technically the lawyers may have been right, but morally I believe it was a rotten idea. What about the citizens in the area who may use that water for drinking and showering? **Does only the corporation count?**

6) Load all the radioactive waste into rockets and shoot them to the sun. Considering it takes roughly 60 pounds of rocket (fuel and vehicle) to send one pound of radioactive waste to the sun and there are tons and tons of radioactive waste, the idea is impractical and wasteful of material.

The long term permanent disposal of radioactive wastes appears to never be calculated into the alleged "cheap cost" of nuclear generated electricity.

If you would like to find out more about this subject of nuclear waste, see the books:

1) **"Radwaste"** A Reporter's Investigation of a Growing Nuclear Menace. by Fred C. Shapiro. © 1981 published by Random House, ISBN #0-394-5ll59-X.

2) **"To Hot To Handle?"** Social and Policy issues in the management of radioactive wastes. Published by Yale University Press, © 1983. ISBN #0-300-02899-7 and ISBN #0-300-02993-4 pbk. Edited by Charles A. Walker, Leroy C. Gould, and Edward J. Woodhouse

3) **"Atomic America"** by Todd Tucker, c 2009, published by Free Press a division of Simon and Schuster, Inc. U.S.A. A history of some early nuclear reactors, problems, and accidents (quite descriptive), as well as efforts to put nuclear power to practical uses.

** **Pavel Alekseevich Cerenkov,** Soviet physicist, 1904-1990. **Cerenkov Radiation,** is emitted when a charged sub-atomic particle such as an electron or proton passes through an optically transparent material (such as air, water, or glass) at a speed greater than light would have in that material. Cerenkov radiation is highly polarized.

The Need For Change To Achieve "Balance"

Fig. #9 is a Graph of the three basic characteristics of a simple harmonic wave of a single frequency.

On this Graph: Displacement is interpreted as the percentage of "order" that is in the "Cosmos". Order in this example being displacement as harmonic wave motion. "Velocity" amplitude is interpreted as the need for change to maintain the harmonic wave. "Acceleration" as the satisfaction of that need for change.

Note Velocity is one fourth cycle out of phase with Displacement and Acceleration.

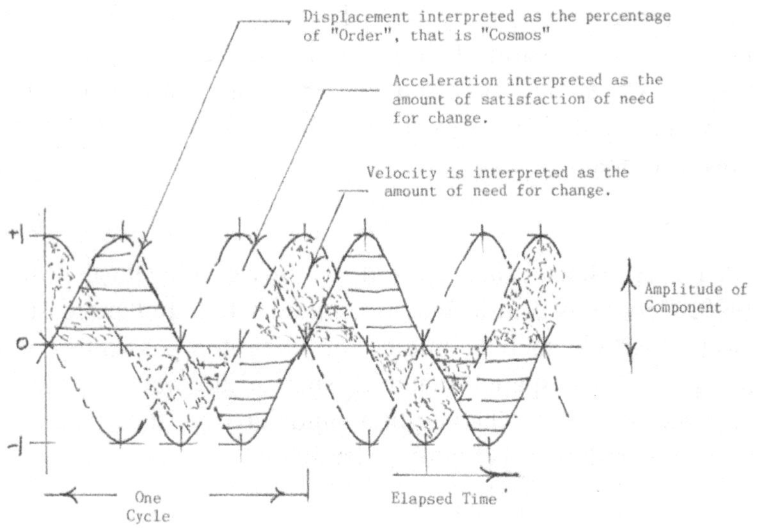

Fig. #9

Things that may never be proven
about the Universe

There are some things that may never be "proven" about the nature of our natural world. "Speculations" on such questions are either accepted by us or rejected. Modern scientists engage in such speculation all the time. Speculations on how big the universe is, it's age, how much matter is in it, is the amount of matter increasing or decreasing, is the motion of our galaxy speeding–up or slowing down, etc.

Any answer is guess work. Scientists use known data and seek trends to arrive at their prediction, their speculative answer.

For my part and in the theory in this book I feel some of the speculations of sages from the past are reasonable, and I have included those speculations in the theory in this book. They are the theory's shadowy outline, and the universe's general character. I present to you now some of these probably never provable speculations:

a) Nothing Comes From Nothing. The belief that there has always existed, a pre-existing "something" out of which the ordered universe of today has been made.

This idea is accepted by many sages, the idea that a "Great Cosmic Ocean" in a state of chaos existed before creation, an "Eternal Reality". "Brahman", an eternal presence. Some say "TAT" existed, that "TAT" is all about us and within us, point anywhere and you will be pointing at "TAT". The Sanskrit word "TAT" means "That" in English.

The Greek **Parmenides** of Elea Greece (5th century B.C.) states that all created things come into being from something that existed prior to their making, or as he put it, "Is could not have come into being because nothing comes from nothing." This may seem obvious, however what is not so obvious is that "Is", the original "Is" must have always existed, it was and always will be "eternal".

b) That there is only One Universe. Universe here meaning, all that exists, ever existed or will come to exist.

c) Universe is the "Totality of all existence". Universe contains all Time, all Energy, and all "Substance". Substance here meaning a substratum a something which, when molded by energy forms the "things" that come and go in our universe such as "matter", "things" that are born, live, and die. That there is, no time or space no substance outside of the universe. If the universe has what might be termed an "outside", an outer boundary it would be space-less and void of time. There is no energy outside of the universe to gain and no energy within it ever leaves, there is no outside for it to go to, so in words a physicist or an engineer might use, the universe is "a lossless reality".

d) Universe is a "continuum", continuous throughout. There are no "voids" within Universe where it is not, no bubbles empty of universe within it. This does not mean there is no change within the universe. The universe has times and places within it where chaos (disorder) may predominate or Cosmos (order) may predominate.

e) The universe oscillates from a state of chaos to a state of cosmos (order). Back and forth, and because of it's immense volume each phase, from being almost total chaos to almost total cosmos, takes billions of billions of years to complete.

f) Change is Constant in the Universe. Change is the most fundamental constant of the basic components characteristics. The three basic components are eternal, their interactions vary but are continuously in progress, or as **Heraclitus** (c.540c-480 B.C. a Greek philosopher of Ephesus) put the situation, **"Everything Flows"**, "Everything is in a state of flux". All things change.

All "Created" or made things in the universe exist in a cycle of change, each "comes into being", is born, lives to maturity, after which each falls into dissolution, that is it looses it's unity, and dies, each lives a duration of time and then is gone.

It is a "Restless" universe we live in, a universe driven by an ability, an energy that can bring about change. **Heraclitus** further believed there is an "Attunement", a "Connectedness" in the universe, all in it is interconnected in that one part not only senses it has a neighbor it also "feels" the condition of that neighbor. This "Attunement" can be compared to the activity of a tensioned string on a musical instrument, which when set in motion transmits it's vibration throughout the instrument and on into the surrounding environment.

Where Does this "Tension" Come From? Heraclitus believed that there is "Strife" between opposites that opposites are continually bumping into one another and affecting one another. In the theory in this book all change in one way or another is "wave motion", waves of displacement of something. In the physical world it is waves of displacement of primordial substance.

In psychology it is "waves of behavior" of "The Self". In metaphysics it is waves of interaction between the goddesses and gods, waves of activity between powers of our world.

g) This is a Universe of Complementary Opposite Pairs. These "pairs" are the extremes of their type what one "Is" what the other "Is Not". Any single thing when divided results in two complementary opposite parts, together or apart the two form the whole of their kind, they are Complementary in that each one of the pair depends on the other to give it existence, and to give one another meaning.

Examples of pairs of complementary opposites:

Chaos and Cosmos (order), Substance and Non-substance, Permanence and Change, Space and Time, Potential Energy and Kinetic Energy, Satisfaction and Need, Female and Male, The Haves and The Have-nots, light and dark, One and Many, Same and Other, Attraction and Repulsion.

The universe is awash with these pairs of complementary opposites, in fact our minds seek out these pairs to make sense of all the data that floods in during our lives.
Dictionary compilers describe one term of such a pair, by telling what it is not, by telling what it's complementary opposite is, knowing the inverse of this is what the thing "is" they are actually trying to describe.

Example: "Light is the absence of dark" and "Dark is the absence of light". A group, is many individuals, an individual, is one of many in a group.

This pairing of complementary opposites also has the advantage, that any condition that is part of one extreme and part of the other extreme can be given a value, an in-between value, Example: "Oh! the dark part of that thing is lighter than that." Put another way, the concept of substance would have little meaning without "non-substance" to compare it to.

h) Only the existence of the three basic components of the universe are constantly in existence. Their interactions vary but their presence is constant.

i) There are three "Qualities" existence can have. The Indo-Aryan-Hindu sages term these qualities "**Gunas**"

Two of these Gunas form a pair of complementary opposites, one is aggressive, the other reactive. The reactive guna is termed "lazy" or "subservient", or "a hinderer of change" It's name is "**Tamas**". Hindus term the active Guna "**Rajas**", it is seen as "restless", "instigative". Chinese sages recognized these two "qualities" in their Yin and Yang. The "Yin" the dark side of the mountain the reactive quality, and "Yang" the light side of the mountain, the warm restless side.

In life there is always a bit of one with the other. Black is accompanied by it's complementary opposite white, and likewise a white thing is accompanied by darkness. The darkness gives value to the light, the light gives value to the dark. If there were only light one would not realize what is present, there needs to be it's opposite to compare it to.

The third Guna is termed by the Hindus "**Sattva**", it is the quality of "Harmony", mutual toleration, mutual need for one another, a sharing of existence, coexistence.

The Identity of the Three Basic Components
of the Universe

In this theory they are termed:

> **a) Primordial Substance**
> **b) Primordial Connectedness**
> **c) Primordial Energy**

In the field of field of Mythology they are the "Body", "Soul", and "Spirit".

In the field of Psychology they are "Body", "Consciousness", and "Will".

In the field of Metaphysics I would be guessing but the body is "The Image", the connectedness is the "Angel" as a swift messenger and the Energy would be termed "The Likeness".

Primordial Substance is the substratum, the quasi-substance that is molded by energy to form all the "created" things in the universe.

Primordial Energy is the power to change substance, to mold it into things, the power to drive substance into waves of displacement, such as electrons, atoms, animals, plants, suns and moons and planets. The interaction of "Primordial Substance" and "Primordial Energy" brings all these "things" about. Each interaction is not eternal, each created wave or thing has a life-span a duration of existence. Energy is the abilities of "Will", "Wisdom", and "Action" (change).

Substance in this theory is itself, inanimate, dead. Energy (the spirit) imparts animation to substance, imparts the power of life into substance.

Primordial Connectedness, The Messenger, is viewed as the carrier of activity, the carrier of the message of life, of breath. In Hebrew mythology, blood is seen as the carrier of the life of the animal, and there is a prohibition against the eating of blood as it is like eating the "life" of the animal. Breath is a sign that the power of the spirit is in the person and that they live. When the spirit leaves a person, their breath ceases and they die.

Let us now move on to the general characteristics of the the universe.

The Tangible Universe

We live in a tangible world, a world of things capable of being touched.

There is "No Action At A Distance"

Here in this world "things" touch "things". This is even true of the thinking process where, sensations touch sensations in the mind and form ideas then, ideas touch ideas and form concepts.

The Ancient Greek philosophers believed there were four basic things that could touch one another in this world, they termed them "essences" or "elements".

Fire and Earth, Air and Water

plus one additional "element", which existed in the world of the gods, it was the substance that the moons, and planets, the sun and stars traveled through. They termed that element of the heavens **"The Aether"**, **"The Quintessence"**, **"The Fifth Element"**.

Men breathed the lower air, and the gods breathed the upper air, the air of the heavens, the air with brightness, and clearness.

These five essences or elements, the Greek philosophers believed made up the tangible universe of men and gods. As the Greek Plato said of the four elements of the world of men,

Nothing is visible without fire
Nothing is solid without earth

and these two are bonded together with air and water. In Plato's book "The Timaeus" he alludes only once to the fifth element Aether saying it is "the brightest part of air". Of the relationship of "the mean", air and water, to fire and earth, he writes, fire is to air, as air is to water, air is to water as water is to earth, these are bound together to form the visible tangible heaven. The proportions of these four elements harmonize the world, the perfect animal, the living universe.

These five "elements" or "essences" are all in contact with one another, touching one another. **There is no action at a distance.**

The Aether

It was out of myth (stories) of gods and goddesses that "The Aether", the fifth element was adopted by scientists, the reason being to avoid action at a distance by providing a pseudo-substance, a medium to directly transmit, that is to carry the vibrations of electromagnetic waves and the forces of electrostatic, magnetic and gravitational fields.

Each theorist gave their version of the properties this "Aether" substance should have. Some thought it needed to possess **elasticity** and **inertia**, some thought it needed rigidity and an elasticity greater than steel yet be able to allow huge planets to pass through it unhindered. Some saw their "Aether" as an elastic solid. Others saw the aether as a gel. Some theorized that their aether needed to be a massless substance.

The Aether Substance needed to be transparent to electromagnetic waves as stars millions of miles away could plainly be seen through it.

The substance needed to carry the forces of gravity through space from one material body to the next, it needed to be elastic, to carry wave motion, and have inertia to transfer Newton's "Innate Force". They called their mythical substance "The Aether" (Sometimes spelled Either).

A list of thinkers that believed the "Aether substance" must exist is quite impressive, all believed the Aether filled all space to the ends of the Universe. Some had slightly different theories as to how it carried the forces. Some saw **vortices **** as carrying the forces, others believed **wave motion carried the forces.** Even though experimentalists could not positively prove the existence of this "Aether", these persons all thought, <u>it must be there.</u>

Here is a list a few of these philosophers and scientists who believed a substance or a pseudo-substance must exist:

Appolonius of Tyana (1st century A.D. a neo-pythagorean sage from Tyana, which in modern Turkey, is Bor), Appolonius believed in the existence of the "Aether".

Aristotle (Greek philosopher, 382-322 B.C.) Believe there was a substance he termed **"prime matter"** that when impressed with "form" became one of the four "elements" (fire, air, earth, water) and they in turn in various proportions constituted all material bodies. This "prime matter" filled the plenum of space meaning the entire volume of the universe.

Johannes Kepler (1571-1630, German astronomer) called the substance that pervaded the cosmos "The ether". He says as much in his work titled "The Epitome of Copernican astronomy, IV".* that nothing except the ether is common to the spheres (the planets etc.) and to the intervals between them. That this ether will not impede the passage of bodies or of light.

Descrates (Rene Des Cartes, or in Latin Renatus Cartesius 1596-1650, French Philosopher) agreed with Kepler that there must be a "something" a "substance" to transmit the forces and pressures of nature from one body to another, an ether that filled the plenum of space.

Isaac Newton (1642-1727), English physicist believed there must be an ether to carry the forces of gravity.

James Clerk Maxwell (1831-1897, British mathematical physicist of Scottish decent) in his brilliant mathematical electromagnetic wave theory, believed, an ether was necessary to carry the waves and fields of force.

Thomas Young (English physicist, 1773-1829) wave theory of light.

William Thomson Kelvin (British physicist 1824-1907). Kelvi devised a dynamical theory of heat, and a theory of light.

Sir Oliver Lodge (1851-1940, English physicist) tried dearly to detect the ether of space, but could not, yet he believed it was there.

Hendrik Antoon Lorentz (Dutch physicist, 1853-1928) tried to form a consistent theory of light-electricity and magnetism.

Nikola Tesla (U.S. Experimentor and Inventor 1856-1943,). Inventor of radio, the split-phase induction motor, the brushless poly-phase induction motor, Various generators, motors and transformers for multiphase electric power generation and use. Wireless electric power transmission, generation of longitudinal electric waves, inventor of a bladeless turbine, and more.

Dr. Wilhelm Reich (Austrian by birth, psychiatrist, psychoanalyst, scientist, 1897-1918) Orgone Energy.

Dr. Harold Aspden (British electro-physicist, inventor, engineer, b. 1928) The Quon-Hadronic Aether.

Ronald R. Hatch (U.S.A. b. 1938, Global Positioning Scientist) "An Ether Gauge Theory", and "Gravitation: Revising both Einstein and Newton".

Gabriel LaFreniere (Canadian, physical theorist) in his wave theory of matter "Matter is Made of Waves" believed the waves were, waves in the either. *****

Albert Einstein (German-Swiss-U.S.A. 1879-1955), made a big noise about, there being no aether, but later in life said it was unthinkable for the aether not to exist.

There were differences in exactly the way the aether delivered the forces of nature in the above person's theories, but all held fast that it must be there as did many, many others both professional and armature. Oliver Lodge said of the ether of space that it needed two qualities or characteristics to carry wave motion, Elasticity and Inertia. Elasticity is the ability to rebound if deformed. Lodge said of Inertia, that it was the ability to over-shoot the mark.

Inertia is the ability to resist a change in velocity or direction of motion. Inertia has also been termed "The Kinetic Reaction Force", Isaac Newton recognized this **"Innate Force"** as the cause of what became known as "Inertia". Newton to his credit said he did not know what caused the "Innate Force".

Sir Oliver Lodge believed that all mass, all momentum, all kinetic energy was that of the ether. *** In the theory in this book these are characteristics of "Primordial Substance" that is of space and of time, animated by Primordial Energy.

* Kepler's work: First Book on the Doctrine of the Schemata.

** **A. E. Dolbear** (Amos Emerson Dolbear 1837-1910. professor of Physics and astronomy) spoke of "Vortex Rings" in his 1897 work "The Machinery of the Universe". In this work he compares the characteristics of the aether and matter.

*** If you are interested in the subject of the ether, see also Sir Oliver Lodge's two books on it, they are "Ether and Reality", 1925, and "The Ether of Space" 1909. The copies I have were nicely reproduced by Kessinger Publishing.

**** If you would like more information on Milo Wolff's standing wave electron, see "Exploring the physics of the Unknown Universe" by Milo Wolff, c 1990, Technotran Press, second ed. 1994 ISBN 0-9627787-0-2 page 180. also you might explore his on-line works.

***** The standing wave electron, is also delt with in Gabriel LaFreniere's" work titled" Matter is Made of Waves" on the internet.

The Problem with the Aether

The problem with the Aether was that even though scientists believed in it as common sense said it must exist, no one could definitely prove by experiment that it existed.

Some Scientists have now come to believe the Aether substance is actually, space itself that space acts like a pseudo-substance. That space is not just a container of things, a plenum, but has physical characteristics.

Walter Russel (American Polymath, M.D. 1871-1963) spoke of "**The Fabric of Space**".

Milo Wolff (American U.S.A. mathematical physicist, 1923b-______) came to speculate that the spherical standing waves that make-up the electron, were waves in space itself, that space was not just a container, but a substance of sorts that had characteristics that could carry wave motion, that is displacement. ****

Charles Noonan Vind -writes of "**Fluid Space**", his work is titled "Cosmos as a material fluid (with added material 2002)".

THE THEORY IN THIS BOOK DOES NOT USE "THE AETHER"

As said, the main purpose of the Aether substance <u>was</u> to carry wave motion, much like water carries wave motion. I have substituted "Primordial Substance" and specified it's character in place of the mythical Aether. The Aether and Primordial Substance, cannot be directly compared as the

"Aether" stands alone, whereas In this theory "Primordial Substance" is more complete.

Primordial Substance carries the wave motion, carries the forces of gravity, carries magnetic and electrostatic fields, carries nuclear forces, and electromagnetic waves.

<u>Primordial Substance includes both Space and Time.</u> Space and Time are like two sides of the same coin, One without the other cannot be, **"Space Substance"** gives time a place "to be", **"Time Substance"** gives space a time "to be", a time to exist in.

I believe Primordial Substance when undisturbed, that is un-displaced cannot be detected. Undisturbed space substance is like a quiet pond with no wave motion, a surface with no form, an endless expanse of quietude. Undisturbed time substance is also like a quiet pond, a pond of eternity, a pond of endless existence, endless duration.

But when this Primordial Substance is disturbed by energy, waves of chaos bring the quiet pond to life, bring it's potentiality into actuality, and soon the restless waves of time and space substances awaken to be, the forms and fields of force of our world, to be chaos and cosmos, disorder and order.

Once energy animates primordial substance it is a quiet pond no longer, and we can detect it's disturbed state. When disturbed that is when displaced, primordial substance's potential characteristics become actualized. We sense the forms and fields of force within it, we sense that which is, the reality we exist in. We sense the living universe.

Partial Matrix Diagram of Physical Universe

To illustrate that each corner square term is part of two worlds. The corner square term is shared by two phases of a cycle of change. In Fig. #17 below "Space" shares the worlds of Atma and Sthula. "Time" shares the worlds of Kama and Sthula. See also page 37. A full diagram of Fig. #17 is on pages 220-221.

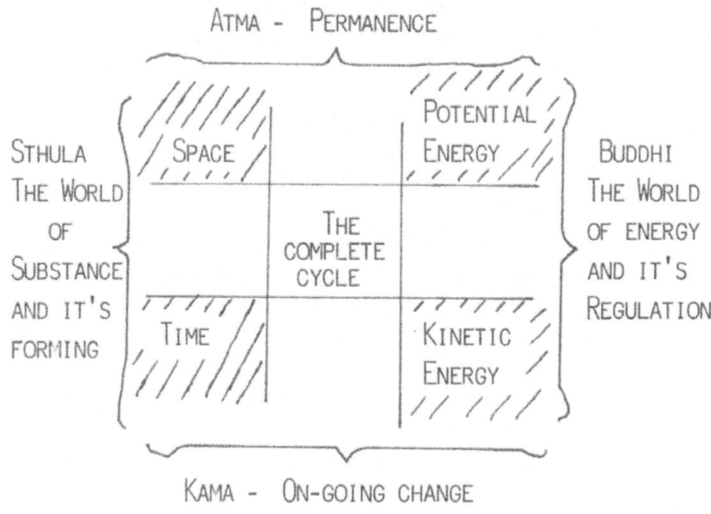

Fig. #17

The Names of "Universe"

Universe, here being is all that exists, has existed and will come to exist, a great unity that is eternal. The Aryan-Hindus believed any name given it could only represent it and none would completely describe it.

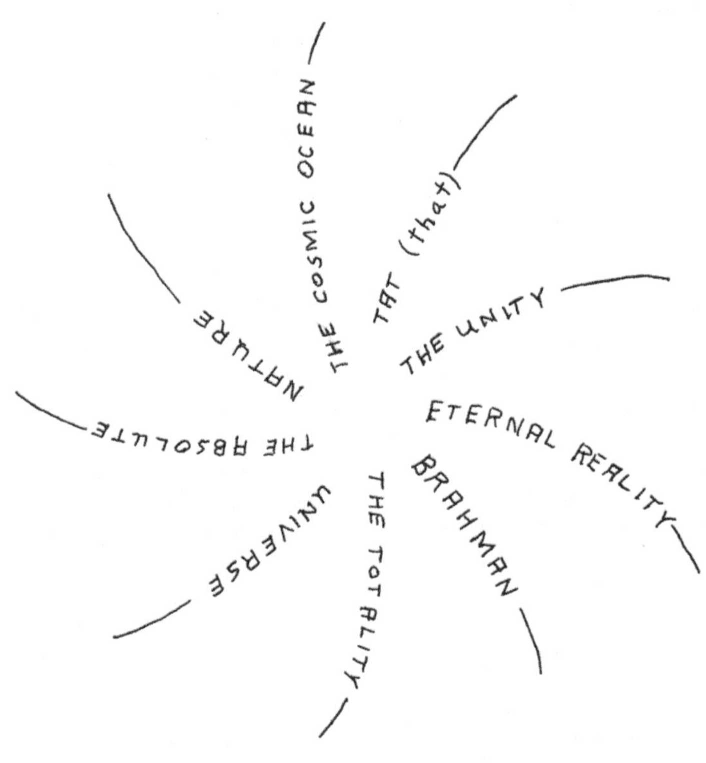

Fig. #10

What Name Shall we give to Eternal Existence?

What name shall we give an existence that had no beginning, exists now, and will never end?

An existence that is a great unity, that encompasses all things, all reality, an existence that contains all space all energy and time itself, an existence that cannot be destroyed. No word can fully describe such an all encompassing reality as the universe, any word or words we choose can only represent it.

The Aryan-Hindus termed that eternal existence "Brahman". Brahman can exist as formlessness or as form. Brahman the eternal, Brahman the imperishable, absolute self contained existence "Universe".

Some of the names given to represent this eternal presence that contains all that exists are:

The Cosmic Ocean, Nature, Universe, The Absolute, Ultimate Being, Brahman, Ultimate Presence, The Ultimate Reality, The Totality, and TAT.

The Sanskrit word TAT is interesting, it's English language equivalent is "That". Point anywhere and you will be pointing at TAT. Even the specific term "Universe" is not the same in all languages, a few examples of it's equivalent in different languages are:

Universe -(English)
Mundus -(Latin)
Weltall -(German)
Universo -(Italian)

Please do not be offended if your native language is not included above, as I am not all that proficient at languages other than English.

The Universe is a grand theater in which the action takes place, and all else needed for the "production". The universe itself exists eternally, it never was not, it is now, and it will always be, it does not "wear-out".

When the Universe is disturbed (that is displaced) it exhibits pairs of complementary opposite characteristics, each a pair of it's kind. Even the state of the universe exists as a pair of complementary opposite states, Chaos and Cosmos, disorder and order.

Modern scientists do not always agree on various aspects of nature, neither did the Aryan-Hindu sages. I have chosen three Aryan-Hindu terms for the three basic components of the universe and their English language equivalent terms. To avoid confusion as to their meaning their definitions are given below as used in this book.

Primordial Substance "Prakriti", the physical substratum of the universe. It is this "substance" that when molded by energy becomes the manifest forms in our universe

Primordial Connectedness, Soul, Consciousness, "Prajna". The connectedness is "<u>The messenger</u>" that delivers the energy into substance to animate substance, to make substance as it were "to live". This "Connectedness" itself, "the soul" is not the message, it exists to carry the message.

Primordial Energy "Prana", is an ability, an ability to mold substance, to animate substance, it is the power to bring about change. Energy in substance is "force" and is motion of substance. There is an energy that drives change, and a second energy, a counter-energy that inhibits change, limits change.

A physicist sees these two energies as "Kinetic" energy and "Potential" energy. See Figure #11, page 82, table of correspondences, in the first vertical column of this table, are the fields of endeavor, the physical universe, The cycle of life of mythology, the self of psychology, and the Aryan-Hindu terms. Each horizontal row lists the names of the three-basic components of the universe in a given field.

Example: **<u>In the field of psychology</u>**, substance is termed **"The Body"**, Connectedness is known as **"Consciousness"**, and energy is termed **"The Will"**. The field of psychology is concerned with the behavior of "The Self". The Self is the center of it's own personal universe, and would be the term in the center square of a psychological matrix diagram.

Anaximander a Milesian philosopher (610 B.C. and still alive in 546 B.C.) believed that there was a non-perceptible substance that filled the universe. That the world of cosmos, the world of order began with the separation of complementary opposites out of this of non-perceptible substance. Anaximander termed this non-perceptible substance **"The Unlimited"**. He believed there was eternal motion in the universe and that when a thing perishes it returns to that out of which it had come, in other words it returns to "the unlimited".

It seems the Aryan-Hindus favored us with a slew of words that begin similarly with the letter 'P', these words are:

Prana, Prakriti, Prajna, and Purusha.

Since different Hindu schools of thought have applied different interpretations to these terms and Buddhists have their interpretations, I will give their definitions as used in this book.

Table of Equivalent terms for the Three Basic Components of the Universe. When the Term "Body" is used it means physical substance or a lump of substance whose pieces act as a unity. Scholars in different fields of endeavor gave different names to the three basic components.

Table of Correspondences of Terms

That Which has the three Components	The Three Components of the Universe		
	That Which is Formed	The Carrier of the "Forming" Forces	The Cause of the "Forming" (the Power)
Universe (physics)	Primordial Substance	Primordial Connectedness	Primordial Energy
Life (Mythology)	Primordial Substance	Primordial Soul	Primordial Spirit
The Self (psychology)	Body	Primordial Consciousness	Primordial Will
Brahman Aryan-Hindu Terms	Prakriti	Prajna	prana

Fig. #11

Let us now take a closer look at the basic components of the Universe as seen in the theory in this book.

Prana and **Prakriti** are the two basic complementary opposite components of the universe. The next two shown below, **Prajna and Purusha**, are perhaps more fuzzy as to their exact meaning, and so their meanings as intended and used in this book are given below.

Prana – Primordial Spirit, Energy, Power, **The Will**. Prana is not life, it is the energy to animate substance to make it to live. Prana is not a substance it is an ability to bring about change.

Prakriti – Primordial Substance. That which is animated by Energy. Energy moves (displaces) substance. We perceive moved substance as forms and fields of force.

Prajna (or Pragna) The Soul, The Consciousness, The Connectedness. Prajna is "The Higher Soul" and "The Lower Soul". A psychologist would term the higher Soul "The Subconscious, and the lower Soul "The Id.

Prajna is the messenger, the carrier of the message of directed energy in substance to animate substance. To a physicist prajna as the higher soul is "Acceleration", and as the lower soul is "Velocity". The carried message is the directedness and amplitude variations of the energy. In metaphysics one has the higher self (the carrying of ideals, beliefs truths) and the lower self carrier of bodily concerns).Prajna itself is eternal.

Purusha – The Man, The Self, or The Person, **Purusha is the message energy sends into substance**. The soul carries this message into primordial substance. There are two halves to this message the higher soul carries a reaction to the message carried by the lower soul. In the higher mind,

Purusha are those things believed to be true. The lower Purusha messages are those of desires, needs, and urges. Purusha is the message of change that Prajna, the soul carries. Prajna the soul in a manor of speaking witnesses the message it carries, witnesses Purusha. Purusha is "The Self", in the sense that it is the person represented in their on-going directed behavior, their goal seeking effort. Purusha is not eternal, behavior dies with the person. Purusha's message however can influence others and as such survive in others.

The Nature of Primordial Substance

Joubert (Joseph Joubert, 1754-1824, French philosopher) commented that:

"Space is the Stature of God".

Space, is the place that is the vastness of God. Space is the potentiality of existence. Space is one half of "Primordial Substance", the other half is "Time". Time is the actuality of existence. The flower people of the 1960's would greet a person with "What's happening man?" What is happening is on-going change, Time is happening. "Man that's heavy." *

* Heavy, as used here means serious or important. Writing this brings back memories of parts of scenes from the 1960's along with the smells of the incense of the sixties, the lingering, filling pleasurable scents. An era of young people just wanting to live and not be dragged into the horrors of war. Those that oppose this view manufacture their own enemy.

The Nature of Primordial Substance

As said, Primordial Substance is that which underlies all manifestations in our universe. From this "substratum" is formed all that physically comes into existence in the universe and when those things formed from this substratum are "un-formed" only the substratum remains of them. By analogy: This substratum behaves like a lump of moist clay. Energy "molds" this lump, gives it shape, and characteristics. If that molding energy were withdrawn, the clay would once again only be a "lump". As a lump it would only have the potential to become "molded" that is "formed". So when Primordial Substance is undisturbed, that is "Un-displaced" it has no characteristics that are apparent, it is un-detectible to us. It appears to be not there. When disturbed that is "displaced" the things and phenomena of our world, of our universe appear and become detectible.

Primordial Substance is "eternal", everlasting. Primordial substance always was, is and will be, it cannot be destroyed, it does not "wear out". Our universe being a unity, like a huge spherical volume, contains all primordial substance, none can wander out of this spherical volume, no space is outside of this spherical volume (if it has an outside). Universe always contains the same amount of Primordial Substance, Primordial Connectedness, and Primordial Energy. None of these three is ever "lost", or reduced in amount.

Let us now look at what Primordial Substance is in the theory in this book.

Space in This Theory

In this theory "Space" is one of the two characteristics of Primordial substance. The space of this theory has three dimensions of extension. To visualize space imagine, the volume inside a hollow sphere. The sphere of the universe is like that, only vast of an almost limitless volume, limitless dimension.

Some see space as "emptiness", as a container of things, but in this theory it is more, much more, it is that out of which, all things are molded by energy.

When not acted on by energy, space is transparent to us. A material object could move through a volume of space that was not affected, that is not displaced by energy only with great effort. If however wave activity, waves of space displacement are already taking place in the space <u>there will be a resistance to the wave motion that is proportional to the level of displacement present</u>. More displacement means greater strain in the displaced substance, greater strain means faster wave transmission speed.

As an example take light waves, they travel very fast, indicating a high strain level. This high strain level may be due to a long period "universe wave" being present in the space the light is traveling through.

So what appears to be a vast emptiness to some, in this theory is a substance filling the vastness of the universe.

Absolute Space

The space of this theory is "absolute" meaning it is unaffected in dimension by what is going on in it. Absolute space is uncompressible, and un-expandable, with no voids in it where it is not, a space continuum. Absolute space to a mathematician or engineer is a place all motion in the universe can be referenced to meaning a grid of parallel lines is identically distant any where in it. Absolute space is the substance of, the "Cosmic Ocean" of the Sumerians, the Assyrians, the Babylonians, and the Indo Aryan-Hindus.

But the Absolute space of this theory probably has a feature the ancient sages did not describe and that feature is, in this theory **there are two volumes** of space existing simultaneously in the same volume. One space sphere is moveable, the other space sphere un-moveable.

Now there these two great space spheres sit, in perfect alignment, within as it were one another's volume. Their volumes occupying one another's space exactly with no place their outer surfaces do not match perfectly. Is there a sort of motion possible that will not move one sphere where the other is not? A motion that will not move their exteriors out of alignment anywhere along their outer surfaces? Now remember the substance of these spheres is incompressible and not rarefiable.

Yes! there is a type of motion that can take place and that sort of motion is a twisting motion, a torque, a brief rotation. The moveable sphere is **rotated in place** compared to the unmovable space sphere and their exteriors will still be perfectly aligned, This is also true for any smaller volume within the huge universe sphere.

When one space sphere is "Torqued", that is rotated compared to the other it is said to be **"Displaced"**. This displacing may be the Torqueing of the entire moveable space sphere of the cosmic ocean, or the "torqueing" of a smaller volume within it.

Torsion Waves

The waves of displacement of "Primordial Substance" in this theory are waves of twist, of torque, they are **"Torsion Waves"**. Torsion waves can radiate outward from their source.

The Price of Displacement

When Kinetic Energy as velocity **"Stress forces"** displace the movable primordial substance **"Strain forces"** are brought into existence in that substance. This "strain" is perhaps the most basic form of "Potential Energy", "Stored Energy".

The more the moveable sphere of substance is "Displaced" by twisting it, the greater the level of strain in it becomes. By analogy **these "Strain forces" can be compared to the "Elastic Forces"** in a compressed steel coil spring, the more the spring is compressed the greater the elastic force becomes in the steel of the spring. The greater the strain forces become the greater is the intention of the compressed spring to return to it's un-compressed state. This is the same with strain forces in displaced primordial substance. Strained primordial substance longs to return to it's un-displaced position. primordial substance which is not displaced has no strain in it.

Unstrained space just lays there like a lox, * like an un-tensioned violin string or a length of rope laying on the floor. **"Stress Forces"** of kinetic energy **drive the space into displacement**, as displacement increases in substance, "strain forces" build (increase) within substance, they store the energy of displacement until the strain forces are unleashed by a reduction or removal of stress forces.

What would the "Strain" in space
do for a space traveler?

A space traveler moving through highly strained space would by analogy be like a sound wave moving through steel, the traveler could accelerate to great speeds. A space traveler moving through space with almost no displacement and therefore almost no "Strain Forces" in it, would find moving through it like a sound wave moving through an un-tensioned rubber band. The traveler's ship would move as if it were in viscous glue.

In the terminology of the physicist, unstrained space would have a low permittivity to motion and a high reluctance for any motion to take place in it. Stressing space would increase it's "permittivity" and lower it's reluctance to motion.

Electromagnetic waves which in this theory are torsion waves, would move through highly stressed space at a faster rate than through un stresses space. the maximum practical velocity would be set by the "strain level" of the space the waves are traveling through. Light waves in our universe appear to have a constant velocity only because over vast distances of interstellar space the average strain level appears relatively constant.

A displacement of space that moves over vast distances as would a "Universe Wave" would have a very long wavelength, and an extremely low frequency, the lowest in the universe, a musician would say such a wave would have a low "Fundamental note". ** Such a "Universe Wave" would undulate at an almost imperceptible rate it would be over billions of years, and so a space traveler or light wave would find it's amplitude of displacement and strain due to it on average, an almost constant value.

It would be interesting if some experimental scientist, a modern Michael Faraday, would determine how great the level in stain of our space is today compared to zero strain. Also what is the range of strain levels within a standing electron wave?

Since "Torsion Waves" have directionality to their twist, other waves traveling through them may be affected by that directionality, it will add to, or subtract from each half of their wavelength. They will suffer a sort of polarization.

This polarization in interstellar space may have already been detected. See the work of **Borge Nodland** and **J.P. Ralston**, also that of **A. Dobado** and **A.L. Maroto** (1997) on the anisotropy of space as concerns the propagation of electromagnetic waves. You may find some of their work on the internet.

* A lox (Yiddish - laks) Smoked salmon, usually served sliced at room temperature.

** The fundamental note, the lowest note for natural resonance, would have a wavelength that depended on the size of the universe sphere.

Future Space travelers I believe may one day map the strain levels in the universe. Then a space traveler will be able to ride the different levels of strain like a riding a hilly landscape. The space traveler may choose to go around some of the strain valleys.

The strain level space map will be a three dimensional map. His situation will be somewhat like that of a mariner wishing to steer around sand bars and islands, some will be seen as permanent and some as having shifting positions or rapidly changing levels. A computer program may navigate these "strain fields" to a distant destination, or the pilot may "wing it" and use manual control.

Before we leave the subject of space there is a question many have asked:

**Is the Volume of the Universe limited
(finite) or is it unlimited (infinite)?**

No matter how expert someone offering an answer to this question is, the answer will be a speculation as no one has been to the boundary of the universe and returned to tell the tale, even light emitted at the boundary at the beginning of creation apparently has not had sufficient time to reach us the universe volume is so vast.

With this in mind here are two persons who ventured a guess.

Lucretius (Titus Lucretius Carus,exact dates not certain 70 B.C.- 54 B.C. a Latin philosopher) believed the whole universe was unbounded (limitless). He was also one who believed "Something cannot be made out of nothing."

Aristotle (Greek philosopher, 384-322 B.C.) in his book "On the Heavens" commented that the heavens (the universe) had no time outside of it that it was eternal, and a unity, a whole and was the only universe that existed. Aristotle believed the heavens (universe), were not infinite. Aristotle's reasoning was that an infinite volume would have no center (and other reasons), see Chapter 7 of book one "On the Heavens", by Aristotle.

It would not matter as far as the theory in this book is concerned if the universe is bounded or unbounded, as long as wave reflection can occur under specific conditions. Reflection is important as it is needed to produce standing waves.

An interesting speculation is, if the universe had a boundary, and a light wave or other wave impacted it, would it be reflected off of that boundary? If a space ship impacted the boundary of the universe, would it also be reflected back off of the boundary, or crash into it?

Persons once searched for the Aether substance, a caution to those who search for "Primordial Substance". Primordial Substance when unaffected by energy is un-displaced and is un-detectible in that condition. However when energy effects this substance it exhibits characteristics, which are detectible.

Note: Matter is made of primordial substance. <u>Matter is a wave phenomena in Primordial Substance.</u> This will be explained in more detail in the section on Matter. Put another way "Matter" is molded or formed primordial substance.

The Aker's "Living Moment"

The Egyptians of the time of the pharaohs * saw time as a period of existence, a duration over which a thing lived it's life. The on-going actual living was done a moment at a time, **"The Living Moment"**, the moment of the "Now". This now moved across a person's life-span like the sun across the sky, between a person's sunrise (birth) and their sunset (death).

The Egyptians represented this view of time symbolically with **the Aker Symbol**, Fig. #12. At the top of the illustration, is the sky, at each end of the sky are the Horizons. The horizons are the limits of the sun's life-span. The Sun is re-born each day at sunrise in the east and dies as it were at sun set in the west. **The two sitting lions** represent, **Yesterday and Tomorrow**, in other words the past and the future.

Between the lions suspended beneath the sky is the **Aten**, the disc of the sun. In some of these Aker drawings below the sun is a symbol termed **"The Anx"**. The Anx symbol consists of what appears to be a stylized standing man, in other words a living man. The living man is standing on a small representation of the earth between the sunrise and sunset of his life, between the horizons of his life. In other words the man exists a moment at a time out of all the moments that span his life, he exists in **"The Living Moment"**, the moment of **"The Now"**.

* The time of the pharaohs, is here seen as from about 2686 B.C. when **Loser**'s step pyramid was built, to 1300 B.C. the time that **Khendjer**'s pyramid was built.

The Aker
Showing the "Living Moment"

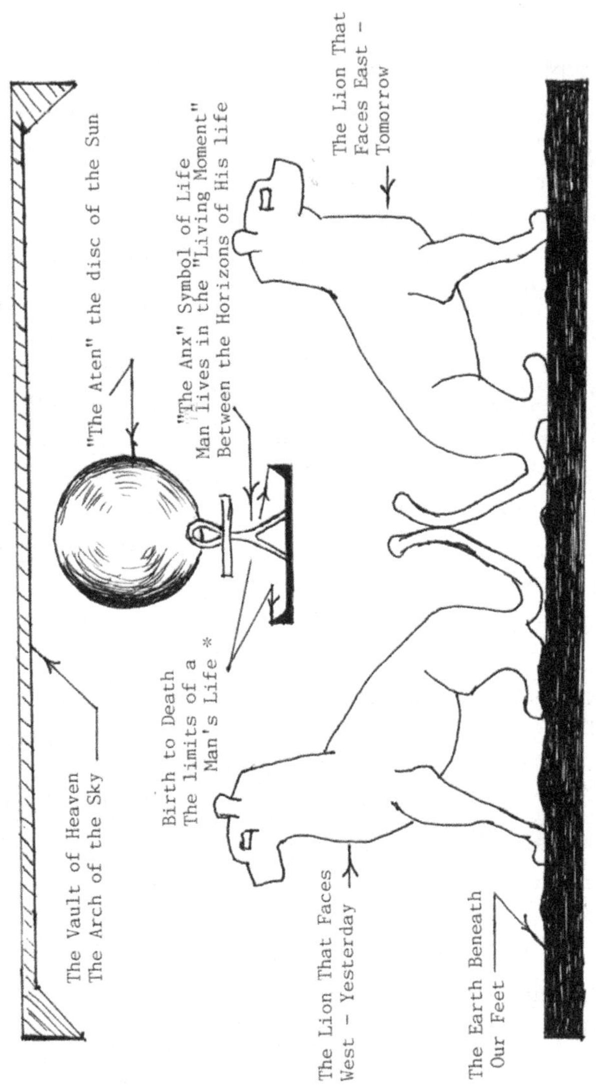

Fig. #12

The Nature of Time

Time the facilitator of change. Time is the persistence of reality. Time lets things be born, lets things have a life-span a duration, and lets them fall into dissolution, that is die. Time removes the old and makes way for the new. Time is the Great Equalizer.

The Roman poet **Ovid (Publius, Ovidius Naso,** 43 B.C. 8 B.C.) commented on the nature of time

"Tempus edax rerum" (Latin)

"Time the Devourer of Things" (English)

For some, time is opportunity and for others time is the gateway to disaster. A period of time is termed a **"Duration"**, it is an elapse of time. All periods of time are part of one unified whole, a "Time Continuum".

"Eternity" is an unending period of time, it is in effect all time, it is the totality of the Continuum.

The Greek philosopher **Aristotle** (384-322 B.C.) believed time was a continuum, he said that time was not motion but the number of motion. Motion is a duration of time over which the level of space displacement changes.

The theory in this book views time as a continuum. The Greek philosopher **Plato** (428/427 to 348/347 B.C.) in his work the "Timaeus" said of time, that time was the moving image of eternity. The Aryan-Hindu sages also believed that time, was eternal. Eternal time being time without limitation to it's duration. Time like Space was not born and will not die, it cannot be destroyed. There are no voids in the universe where time and space do not exist. Both time and space in this theory are "substances" each in it's way is the complement of the other, they are the extremes of a pair of opposites, one is what the other is not. Time is all actuality. Space is all potentiality.

No change in space can occur without time, no interaction with space and energy can occur without time. Space gives things a "place" to be, Time gives things a "Time" to be, a "now" for them to exist in.

Time and Change

The Greek philosopher **Heraclitus** (c540c - 480c B.C.) commented that all things are in a state of "flux", that is in a state of "flow", continual change. Time facilitates that flow. Without time ours would be a frozen world.

By way of analogy our world would be like a series of still photos on a strip of movie film with no way to get from one photo to the next to show the action if time were not present. Another analogy of time is that of a flowing river which is never the same from moment to moment.

The English poet **Alfred Tennyson** (1809-1892) put it this way:

"The old order changeth, yielding place to the new."

With the new, there is opportunity, a chance for something different. Time gives "the new" a place to be where "the old" once existed, it gives the new a "time to be". Time makes possible the changing seasons, the motion of the planets, the flight of birds, and life itself.

The Egyptians of the time of the pharaohs believed we are actually living only a moment at a time, "The Living Moment", the moment of the "Now". Though the ancient Egyptian explanation was expressed symbolically, a scientist of today essentially has an identical view of the situation.

I would say the Past and Future are Potential-ities. The past is a proven potentiality, consisting of things that once were actualities. The future is a time of what might be possible, a time of what might one day be actualized, an extrapolation, an extension of what further combinations of conditions or past or present events might lead to.

"The Now" is very special, we know it is there or we could not exist in it, however it is so fleeting, so brief it is a wonder that the moment of "The Now" exists at all.

"The Now" is actualized time. "<u>The Now" is the actualized moment</u>. "The Now" is the "Living Moment" the moment each thing exists in, the instant of time it is alive in.

Time as eternity is potential time except for the moment of "The Now". The moment of the "Now" travels across eternity, travels across potential time, like the sun across the sky, and just as earth has only one sun to call it's own, eternity has only one "Moment of the Now". There is only one "Now" sweeping across eternity. Recall, in this theory, time, and space, and energy are a unity. Eternity is the time of that unity over it's whole existence, "The Now" is the actualized instant out of an eternity of potentiality.

How Long is "The Now"?

Aristotle the Greek philosopher (384-322 B.C.) commented on "The Now" the present moment. Aristotle wondered how "The Now" could have time to exist at all. Aristotle reasoned that if the moment that is "The Now" had an extended duration the present moment of the now would include parts of the past, and parts of the future. To avoid this difficulty, the now must have an exceedingly short duration, almost so short, as to not be there at all.

The Roman poet **Virgil (Publius Vergillius Maro**, 70-19 B.C.) commented:

"Tempus edas rerum" (Latin)

"Time Flies" (English)

However short a moment of "The Now" is, we experience and exist in a series of them during the duration of our lives. So how does the moment of "The Now" seem to exist longer for us than a microscopically short instant?

I believe experiencing the now is similar to **"The Persistence of Vision"** in humans.

Our perception of a light wave impacting the optic nerve takes place during an extremely brief time, yet the impact of many consecutive waves appears to be continuous. This is due to the way our mind's nervous system functions. The impression made during an instant of the now "persists" for a brief while so that a new instant of the now blends in with the impression of the old fading "now". I believe this is so with our impression of time itself. In our mind experiencing the impression of the last "past now" lingers in our memory and the new moment of the "now" is impressed on it.

The result is that time seems to flow continuously, from now, to now, to now, to now. Thus "The now" of the present moment, is ever present during our lives. The "Living Moment" lives on.

With the above explanation of "The Now", we can ask:

Do the past and future exist at all, or is only "The Now" existing?

Some persons have inferred that only the present moment of "the now" exists.

Time is continuous, a continuum, an eternal reality. Out of this time continuum come all the moments of "The Now", like drops of water from a cosmic ocean. All time, as eternity, is a potentiality, except the present moment, The Moment of "The Now", that moment is an actuality. Put another way, Eternity is the potential for an unending series of "Nows" to actualize out of. **Both Potential time and actual time are real.** Actualized time is like actualized space, in that actualized space is the space we occupy with our bodies this moment. Potential time is the potential moments our bodies do not as yet exist in. Potential space is the space our bodies have the potential to exist in, but at the moment do not.

Does Time Have a direction to it?

Well! No and Yes, which depends on the sort of time we are speaking of, eternal time or a in a duration of time a "thing" exists in.

Eternal Time, has no direction, no beginning, no end, only existence, only presence as potentiality. A "thing" with a beginning and end with a duration in-between does have a direction to the series of moments it exists in.

For a thing with a duration, a life span, time flows only one way. Usually the direction is seen as flowing from the thing's beginning to it's end, and then the thing exists no more as a unity. For things that are born, their time is often ephemeral, fleeting life-span compared to eternity. To put this situation more poetically, Our existence passes like a shadow. **See footnote page 104**.

The Time in this theory

In this theory Time is a substance. Time substance is similar to space substance, in that both have two parts to them, each has an un-moveable half, and a moveable half, both halves of time occupy the same volume of space. When the moveable half is moved out of alignment compared to the unmovable half it is said to be "displaced". When time and space substances are displaced a strain comes into existence in them, an elastic strain that tends to return each to the un-displaced condition once the displacing forces are reduced or removed.

Eternal time is un-displaced time. Once eternal time is displaced, the displaced time becomes actualized time. The displaced portion of eternal time then has a "duration" to it, and a "strain" in it. Displaced time has a life-span to it's presence, during which it's allotted series of potential moments, tick, tick, away, **only the actualized moment exists**. Tick, tick the actualized moment moves across eternity, at the end of the actualized moments allotted duration it sinks back into eternal time, and is gone, sunk back into un-nows, the moments no "thing" lives actualized in, eternal moments.

Like displacing space, displacing time is like stretching a rubber band, elastic forces in the substance of time appear that tend to return it back to it's un-displaced condition. The greater the displacement, the greater the elastic strain forces become, forces that urge return to a state of un-displacement.

The Cycle of Existence

This displacing and un-displacing then forms the duration of a thing, it's cycle of existence. The length of the duration that is the life-span of a thing is the tricky part. I believe it is set by the propagation time of the displacement, which is set by the level of strain in substance. More strain, faster propagation rate, no strain, no propagation rate, and eternity.

Since this theory is a wave theory, in our universe we have waves of displacement passing simultaneously through the same volume of space, with other waves of displacement also present. It is a good thing the elastic strain forces do not wear-out, or the continuum of time would suffer "work hardening" and crack, with all the flexing going-on, all the straining and un-straining.

A science fiction writer might pick-up on this idea of "A crack in time", what would the crack consist of? Perhaps unstrained time, eternal time in the zone of space in the area of the crack. It would be a sort of "Time Fault Zone" in space.
I am not saying such a phenomena exists, just imagining "what if".

Space and Time Substance are bonded together, Where there is one there is also the other, when one is "displaced" so is the other. By analogy they are like the front and back of a coin in their existing together. In that way time controls the duration, and space displacement controls the character and extent (form) of born "things". Some questions are still unanswered here such as is the strain in space and the strain in time of equal amplitude? Are the two strains in reality one strain? or are they independent of one another. Both the time and the space substances of this theory fill the universe, there are no voids in them, no place in the universe without them. Outside the universe if it has an outside, there is no space, no time, and no energy.

How is Time classified by those in the know?

Time is classified by those in the know by how it is perceived as follows:

Absolute Time - Absolute Time is the time that exists independent and unaffected by what is occurring in it, it's dimensions remain unchanged even within an object traveling at light speed and beyond. Absolute time ticks evenly, a minute, is a minute is a minute, any time, any place.
The time of Newton and of this Theory, is "Absolute Time".

Subjective Time - Subjective time is time as we "feel" it passing, our experience of events or non-events tends to let one "feel" the passing of time "going faster or slower". Subjective time is not stop-watch time, it is said to be due to one's ego, by the consciousness of a person. An example of Subjective Time is: "It seems to be taking her a long time."

Objective Time - Objective time is historical time, it also has been termed "public time". Objective time is the time of a sequence of events in a series, like a "time line" of history, where events are noted along a duration of time.

Example: "A certain event occurred in 1842, and that other event occurred in 1903".

Cicero (Marcus Tullius Cicero 106-43 B.C. Roman * pleder in the law courts of Rome, Cicero was eventually elected as one of the two chief magistrates), Cicero once commented:

"0 Tempora, 0 Mores!" (Latin)

"Oh the Times, Oh the Mores!" (English)

He was talking about "objective time" the "mores" he was speaking of were the fixed customs, or moods of a period for a society or particular group. Mores are also shifting moral attitudes.

What About Time In a Relativity Theory?

In a relativity theory that is based on common sense and not some fantasy about mashing time and space together in some rubbery conglomerate, absolute time does not alter, however the frequency of waves can be altered by the Doppler Effect.

In a relativity theory the frequency of waves appears to change depending on the relative direction and velocity of the source of the waves and that of the receiver of the waves. The Doppler Effect discovered in 1842 by Austrian physicist **Christian J. Doppler** (1803-1853) is a scientifically proven fact. It is the cause of the rise in frequency of the whistle or horn of an approaching train and the lowering of the frequency of sound from a receding train.

The Doppler Effect has now been extended to the workings in the internal structure of an electron's standing wave. See the work of Canadian **Gabriel LaFreniere** and his "Matter is Made of Waves" theory posted on the internet, in 2009 it was still posted.

Can Time be speeded up or slowed down?

Absolute Time, - Never, Absolute time does not shrink or expand speed up or slow down.

Objective time, - Perhaps. The winners of wars write history, the authorities in an organized religion write it's history. Writers are humans, humans are fallible, humans will "Lie like a fish" when they feel it is worth their while to do so. Altering the dates of historical events or leaving them out completely, in it's way this is altering objective time.

Subjective Time, - Is always changing. Subjective time is the passage of time we can "feel". Subjective time is when "time feels like it is crawling by", or when time seems to have "slipped by" (not been noticed ticking away) and we say something like "where did the time go".

Footnote: **On the one way directional flow of time**. If time is viewed from the point of the death of a thing, it also appears to flow one way (from the time of death toward the time of birth of the thing), the opposite direction from time seen from the moment of birth of the thing. After all, for the thing itself there is no time after it dies, just as there is no time for the thing itself before it was born, So **it is a matter of how one cares to interpret the direction of the flow of time.**

John Donne (1572-1631, English poet, priest) wrote

 "Never send to know for whom the
 bell tolls; it tolls for thee."

"Send" here is used in the sense of "ask", seek.

The cosmic heart beat is a tolling that marks off the life of cosmos and the times of our lives. It tolls away our minutes and tolls away our hours, it tolls out our days and our nights, our summers and our winters. Our beginnings come and our beginnings go, all as the soul of cosmos looks on.

John Donne in the quote above inferred, it may be just as well not to ask for whom the bell is tolling, alluding to an old practice of ringing a chapel bell to announce a funeral. In asking one runs the risk of finding out it is announcing one's own end.

Side Lite about Time: A very perceptive someone once commented:

In the past there was much more future than
 there is now.

I give them five stars for that comment, it is very good, and for an individual life span it is true. An individual in a way uses-up their future as they live their life.

The above statement could also be re-stated as:

In the future there will be more past than
 there is now.

The Displacement of Substance

The upper two spheres show no displacement, indicated by the matching positions of the registration marks.

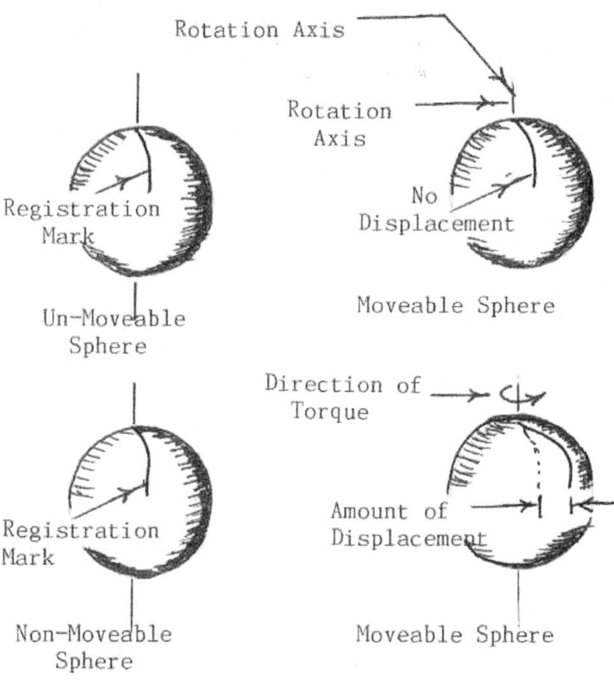

Fig. #13

The lower two spheres show the moveable sphere has been displaced, as the registration mark on it is no longer aligned with the mark on the un-moveable sphere. The moveable sphere has undergone a rotation, a torsion.

Illustration, Displacement of Substance
In This Theory

The displacement of either Time Substance or Space Substance is the same in principle in this theory. In this example the displacement of time will be spoken of.

Time substance consists of two spherical volumes completely filled with that substance, they occupy the same volume of space simultaneously, their exterior surfaces match perfectly. One of these spherical volumes is stationary, the other moveable.

The only displacement motion possible under these conditions is a rotary twist, a torqueing motion.

When time substance is displaced, "duration" appears. When space substance is displaced "strain" appears in it, which we sometimes perceive as a "form" or field of force.

The Way Of Energy

Energy is come from the vasty deep, <u>it is such stuff as dreams are made on,</u> * it seems to appear from nowhere and go back to nowhere, and yet it is there, it is there, it must be there. It is in the Ice and Snow, the Wind and Sea, Earth and Flame, You and, and I.

* The sentence **"it is such stuff as dreams are made on"** is from Shakespeare's play "The Tempest" act 4 scene #1. Shakespeare is referring to the form of a thing, it's behavior, character.

In the 1942 movie "The Maltese Falcon", Humphrey Bogart's character detective Sam Spade referred to the thing itself, when he said **"The stuff that dreams are made of"**.

Bogart's character in the movie not being an English major, apparently saw "stuff" as both the substance and the form substance can be made into as one reality. However technically from the standpoint of using words, Bogart's character in saying **"of"** was referring to "Stuff" as a lump **of** un-formed substance. In other words he referred to a basic quality, the fundamental something of which a thing is made, it's essence and not the form the substance can be made into.

The dreams about the Falcon of the movie were, to those who wished to posses it, wealth, power, perhaps a life of ease.

Stephen Humphrey Bogart, (1899-1957, U.S.A. actor) had a trademark scar and lisp both the result of a WWI shrapnel wound.

The Nature of "Energy"

Energy is an ability, the ability, the power to bring about change. When infused in substance it is the ability to animate substance or to "mold" substance. Energy is not a "substance" is a non-substance. Energy is the complementary opposite of "Primordial Substance". The Aryan-Hindus named the ability of Energy "**Buddhi**", which means "**To Awaken**". Energy is the power to to awaken, that is to animate substance. The power to bring substance to life.

Because of the hidden nature of "Energy" it is mysterious. Understanding of "Energy" came slowly, each field of endeavor, gave energy a different name. One termed energy "The Spirit", another termed it "Buddhi", still another termed energy or power of "The Will", a physicist today terms it "Energy".

In the field of psychology where the behavior of animals including man is studied, Energy is the power that drives behavior, motivation, it is the urges that give behavior direction and priority. Those that speak of the power of the mind, speak of the power of "The Will".

Of the power of "The Will", the following has been said:

> **"They Can Because They Think They Can"**
> **Virgil** -(Publius Vergillius Maro,70-19 B.C.,
> Roman Poet)

> **"No One Knows What He Can Do Until He Tries"**
> **Publilius** -(Publilius Syrus, 1st century Latin
> Writer)

> **"Where Your Will Is Ready, Your Feet Are Light"**
> **G. Herbert** -(George Herbert, 1593-1633,
> English Poet)

The power of a society to bring about change is also spoken of as "The Will of the People". Energy is spoken of in myths as "The Spirit". It is the spirit of deity that brings otherwise dead substance to life, the ability to breath and, to be an animate being. "The Spirit" inhabits the substance of living things, when it is withdrawn by the deity, the living thing dies.

As you may know **"Power"** is the use of energy over a period of time to perform "work", work being the overcoming of a resistance to move an object a distance, In the theory in this book this "moving" is in the form of displacement of the primordial substance of the universe.

James Watt (1736-1819, Scottish engineer) inventor of the modern condensing steam engine, allegedly said to King George the third of England:

> **"I deal in an article kings are said
> to be fond of, Power"**

All motorized vehicles, cars, planes, boats, power-tools, electrical home-appliances, earth moving machines etc. all need energy to move. Wars have been fought over who controls sources of energy such as coal, oil, uranium, hydropower, food etc.

In cold climates the metering out of energy as heat can mean the difference between life and death. But what is this thing "Energy"?

The Greek Philosophers and Energy

Aristotle's **"Unmoved Mover"** could rightly be seen as Energy. Energy is not substance, not space and not time, <u>Energy is an Ability</u>, an ability to bring about change. Energy is the only ability that can bring about change. Aristotle in various places in his works, such as in his "Metaphysics", saw energy as being of two forms, or characteristics.

One form, is on-going change, motion **"Energia"** (activity). The other form is what has been seen as "stored energy", **"Dynamis"** (stored force), in today's terminology "Potential Energy".

So we have energy as Energia, which today is termed "Kinetic Energy" the energy of motion, and Dynamis, energy which is held, "stored".

Under the needed conditions Dynamis can be released to become active energy, kinetic energy. Back and Forth. Back and forth energy of the universe goes exchanged between these two forms or characteristics of the one ability, the ability to bring about change.

Kinetic energy can be seen in the motion of a body, an object, Potential energy can be seen in an object that has been elevated in a gravitational field, in a compressed spring, or in the spherical standing wave structure of electrons, protons, neutrons, or in atoms made up of those sub-atomic particles, those bundles of organized waves.

In today's physics and other science text books:

"Potential Energy" is described as energy possessed by an object due to it's "Position", "Composition", or "State" (that is solid, liquid, gas, or plasma).

Kinetic Energy is that due to the motion of an object, and I believe the motion of the substance of a wave, any sort of wave. In a traveling wave there is energy of motion and in a standing wave their is "Potential Energy" due to a displaced position of space substance or time substance. Potential energy is held in the substance as "strain forces".

The Hidden Force Within An Object

Isaac Newton (1642-1727, English physical scientist and mathematician) *** recognized a hidden force with an object an **"Innate Force"**, others have termed it a "Kinetic Reaction Force", a force that is within atoms or molecules, a force that resists a change of direction, or velocity.

A body with this property of laziness, a resistance to a change of velocity or direction, is said to have **"Inertia"**. In their frustration to explain the "Innate Force" and the Laziness or "Inertia" of an object, modern physicists have said Inertia is a property of "Mass", thereby avoiding saying they do not know the cause of the "Innate Force" or of Inertia.

*** See the "Mathematical Principles of Natural Philosophy" by Isaac Newton, often shortened to "The Principia", The Principles. In the beginning of this work are a few definitions. In definition **#1** "the quantity of matter" is explained, In definition **#2** "the Innate Force" or vis insita, the cause of inertia is spoken of. Newton did not explain inertia only commented on the force that brought it about. It is a wonder "The Principia" was published at all, publication was due to a friend of Newton, astronomer **Edmund Halley** (1656-1742 English), who on getting Newton's permission, personally funded the "Principia's" publication. Halley is the fellow that observed a comet in 1682 now named after him "Halley's Comet", this comet has a very eccentric 76 year cycle orbit around the sun. Halley's Comet should reappear in the years 1986, 2062, and 2138.

I believe Latin was the language Newton wrote the "Principia" in. The English translation of "The Principia" by **A. Motte** appeared in 1729 it is the form most scientists today have come to know it.

In our quest to find the meaning of "Energy" we come to a little mathematical speculation
A. Einstein (1879-1955, of German, Swiss, American citizenships, mathematical theorist) may have borrowed * an equation that purports to show the equivalence of "mass" and "energy". as follows:

$$E = \text{mass times (c squared)}$$

c being the velocity of light in a vacuum. With a little algebraic juggling, if any two values in the equation are known the third value can be found:

$$\text{Mass} = \frac{\text{Energy}}{\text{c squared}}$$

c = the square root of (energy divided by mass)

Mathematically this all appears to be very neat, almost to neat, and in practice, results in unrealistically high values for energy, values simply not found in practice.

To get some semblance of sense out of this formula the work of retired professor of quantum chemistry and nuclear chemistry at the college of science, University of Baghdad Iraq (now living in Amman Jordan), **Mr. Bhajat Razzaq Jaafar Muhyedeen** can be consulted. ** Mr. Muhyedeen asserts that the equation as stated by A. Einstein is defective as the term c2 is a mashed together term comprised of the velocity of momentum and a conversion factor 'c', which converts momentum units into energy units. Muhyedeen asserts that the c of the momentum of un-accelerated charged particles emitted by nuclear reactions is lower than c and that a new constant 'b' which he gives as:

b = 0.624942362 x 10 to the 8th power meters per second

should be used in the formula, to restate the formula we have:

$$E = \textbf{(momentum) x c}$$
note: this is not times 'c' squared

$$E = \textbf{(mass of the charged particle x b) x c}$$

or

$$E = mbc$$

This will result in realistic values in practice. Mr. Muhyedeen then interestingly goes on in his works to use his new '**b**' constant to determine the wavelength and frequency of charged particles emitted from nuclear reactions.

What the equation does tell us is there is energy in the mass of a particle of matter.

Momentum is spoken of in the above formula Momentum in the world of physics has been defined as **"the quantity of motion"**.

Momentum = mass times velocity

* Professor **Bartocci** of the University of Pruegia in Italy explains that an industrialist **01into de Pretto** from Vicenza authored an article that was published in the scientific magazine "Atte" in 1903 (two years before A. Einstein published his "Special Relativity Theory" in 1905 and 12 years before A. Einstein's "General Relativity Theory" was published in 1915.), an article that had to do with the now well known equation.

** Two works of Bahjat R.J. Muhyedeen of interest to us here are:

"On a Heuristic Viewpoints Concerning the Mass, Energy and Light Concepts in Quantum Physics" European Journal of Scientific Research, ISBN 1450-216X Vol.22, No.4 (2008) pp584-601, and from Vol.26, No.2 (2009) pp 161-175 an article titled "New Concept of Mass-Energy Equivalence."

The Universe speaks with one voice
To animate all

The Soul of the Universe,
"The Connectedness"

The Connectedness is the "link" between Primordial Substance and Primordial Energy. The Connectedness lets the two complementary opposites have continuous influence on one another.

Historically and in this theory substance of itself is considered to be inanimate, inactive, dead. The Energy of the power of creation is delivered into substance and substance becomes animate. "The Connectedness" delivers the message of energy within substance and substance then lives.

"The Connectedness" in the fields of medicine and psychology is termed "**Consciousness**".

In the world of mythology the connectedness is termed "**The Soul**", "**The Messenger**". The Soul delivers power to breath, and the power to organize that which can be made to live.

As concerns a person's personal soul, some have inferred a soul can die.

"Let my soul die the death of the just,..."
(Numbers ch.23, vs. 10.

The soul itself is immortal, so how can it die? A soul is seen as dead when it no longer delivers the message of "livingness" into substance, No message, no life. The "Soul" of itself exists with or without carrying a message.

Fig, #14

The Soul, The Ba the messenger. The Ba in
Fig. #14 is represented by a Stork carrying the Egyptian
hieroglyphic symbol for the Ka, two upraised forearms. The
Ka is here seen as the activity of the Ba. The Ka is the
message carried by the Soul.

The Egyptians at the time of the pharaohs believed, when the
Ba left a person their body died, the Ba itself was immortal.
The Egyptians believed the Ka did not die when the body
died, if the person had not lived a worthy life, the Ka was
transferred to a newly born person and they had to live and
an earthly life of struggle all over again. This is consistent
with the theory in this book, in it the Ka is the message of
life, The Ba, is the Soul, the Connectedness, the messenger.

The Names of The Soul
and
of it's three aspects

From antiquity some have seen the soul as having three parts. The one soul spoken of by the Greek Philosopher Plato and by later writer(s) of the Kabala had three aspects. The Soul as a Unity is here termed: **"The World Soul"**, **"Primordial Soul"**, or **"Universal Soul"**. The three aspects of this Primordial Soul are:

The Divine Soul, **Nous** (Intellect), Neshamah (breath of life). The activity of this characteristic of the soul is a reaction to conditions, a reaction to inhibit otherwise uninhibited change.

The Human Soul, The **Epithumia** (Appetite), Rauch (the breathing of life is it's activity), and The Waking Consciousness.

The Animal Soul, Thumos (Passion, Angler, Fear, Desire), Nepesh (or Nepes, the ability to act and feel).

Fig. #15 is a table of correspondences of the above three aspects of the soul.

In the theory in this book the soul is also termed "Consciousness". Consciousness is our "self awareness". Psychologists theorize the Consciousness has three parts, The Sub-Conscious, The Waking Conscious, and The ID which is also termed, The Un-Conscious.

See glossary pages 241, 245, 257, 259 for definitions of these three aspects of consciousness.

Table of Correspondences of the three basic characteristics
of the one Soul, The World Soul

English Term	Plato's Terms	Hebrew Terms	Psychological Terms	Physical Terms
The Divine Soul	Nous (Intellect)	Neshmah (Breath, Life)	The Subconscious	Reaction
The Human Soul	Epithumia (Appetite)	Rauch (The Breathing of Life)	The Waking Consciousness	Resolution of Complementary opposites
The Animal Soul	Thumos (Passion, Anger, Fear, Desire)	Nepesh (or Nepes) (The Ability to act and feel)	The Id	Action On-going Change Directed effort

Fig. #15

In this Matrix Diagram

Note: not shown in the diagram are the God Ra and the Ka. The Ka in this theory is the activity of the Ba, the soul. The activity of the soul delivers a force, a hidden force of change, that force is "Ra". Ra is opposed by a force of limitation represented here by the Goddess Sekmet.

Author's speculations on the correspondences between the Egyptian Goddesses and Gods and the Aryan-Hindu terms are indicated.

The corner squares of the nine square matrix diagram are shared by two adjacent sides (each side consists of three squares).

The Goddesses And Gods Of The Time Of The Pharaohs

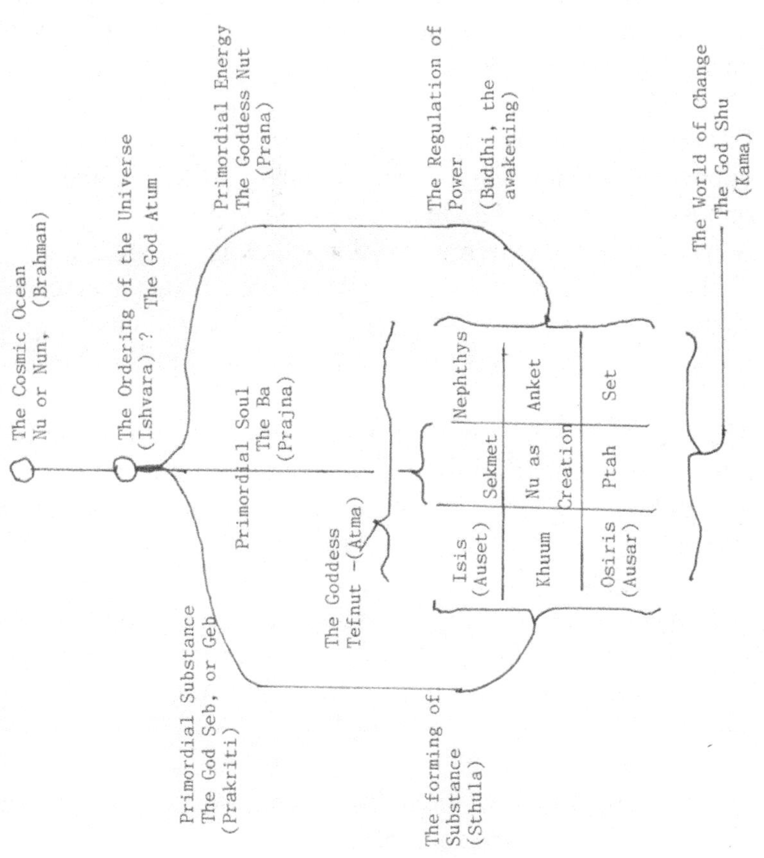

Fig. #16

All Three Basic Components Together

The two basic components of the universe proposed by the ancients have now been described, substance and energy, and a third basic component has been added to allow interaction between substance and energy. **See Fig. #16, page 121.**

Energy and Substance are a pair of complementary opposites, each having what the other does not, they are opposite extremes of a kind.
Substance is one extreme, it is that which is changed, Energy is the other extreme. Energy is an ability that can cause change. it is a non-substance. The third component is the messenger that delivers the power of change into substance.

The message of energy is in two parts. One part is the on-going actualized effort to bring about change, kinetic energy, as **"Velocity"**, active in substance it appears as a **"Stress Force"**.

The other part of energy's message is potential energy (stored energy) it may resist or encourage on-going change. The message of potential energy is **"Acceleration"** in substance it appears as **"Strain Forces"**. Acceleration takes place when a substance undergoes a change in speed or direction of motion, including speed of rotation or the direction of the axis of rotation.

Average   <u>change in velocity</u>
Acceleration = time taken to make change

Acceleration may be from a decreasing speed or an increasing speed. In popular vernacular an acceleration due to decreasing speed is sometimes termed "deceleration". The

universe is never quiet, energy is the cause of the restlessness.

Energy is eternal, it does not increase in it's total amount or decrease. Energy does not wear-out, it is only a question of which part, potential energy or kinetic energy will be the one determining how change proceeds.

<u>In this theory</u>, velocity of substance delivers a force of change, a stress force that moves (that is displaces) substance, that molds substance by the push of stress forces.

As a result of the displacement "strain forces" are generated in substance., they act as stored elastic forces, stored energy forces that tend to return molded (displaced) substance back to it's un-displaced state at the earliest opportunity,

The strain forces put limitation on the amount of displacement substance can achieve due to the push of the stress forces. In so doing fields of force with "shape" result and substance is given "form".

Greater stress force means greater displacement, greater displacement means greater strain forces.
Reducing one reduces the other.

Is Force A Push Or A Pull?

Push forces can appear as a pull. A pull being, a lesser push dominated by a greater push. The resulting motion of a material object in such a force field would appear as if the object was being "pulled" toward the lesser force.

The Cycle of Energy Regulation

The energy in the universe is continually sloshing like water in a trough, sloshing between the "Actuality" of the on-going force driving change and the "Potential" reaction to the on-going change.

Empedocles (c.490-430 B.C., Greek philosopher, poet) appears to have based his view of the universe on his observations of plant and animal life, transferring those observations to the universe as a whole.

Empedocles saw the universe as alternately at rest or in motion. He believed "Love" and "Strife" were the cause of change. To put this another way he believed "Love" and "Strife" made the world go around. **Love makes the one out of many (Love combines). Strife makes the many out of the one (Strife separates).**

Empedocles saw the universe as a vast spherical volume, when love ruled opposites united in harmony, when strife ruled opposites were driven apart. By definition "Strife" is exertion or contention for superiority, a struggle between opposites. Love being a harmonious relationship.

Empedocles also believed there were four "Root Substances", fire, air, earth, water, and it is these that exhibit love or strife with one another.

So If I may interpret Empedocles beliefs as they apply to the theory in this book, He saw life in one of two extreme states, as strife, a struggle "same" versus "other", or alternately as co-operation, "same" harmonious with "other" Sharing existence with "the other" is love, a co-existence between what is potential and what is actual.

*** **"the other"** can be seen one's psychological "Shadow Self", or society as "not self", or as the world outside of "Self",

Considerations on
Energy, Substance and Matter

a) Aristotle (Aristotle Greek philosopher, 434-322 B.C.)in his book "Metaphysics", Book IX, Ch.8 In effect said that, both "substance" and "form" contain energy.

In this theory substance is substance. Form is substance acted upon by energy. Motion in this theory is wave motion of substance. Substance is displaced by kinetic "Stress Forces" of velocity. Substance in a displaced state contains Potential energy as "Strain Forces".

b) A. Bain (Alexander Bain, Scottish philosopher, 1818-1903) in his book "Logic" 1870, believed, matter, force and inertia, are substantially the same force. also, that force and matter are not two things but one thing, and that force, inertia, momentum, and matter are all one fact. (See A. Bain, "Logic" volume II. Longmans Green and Co. London 1870.

c) Frederick Soddy (English, radio-chemist, 1877-1956) in 1904 "Radio-activity: An Elementary Treatise, from the standpoint of the disintegra-tion theory" in the "Electrician" 1904, regarded atomic mass as a function of internal energy.

d) A. E. Dolbear (Amos Dolbear U.S. physics-astronomy professor, 1837-1910) in his work "Matter, Ether and Motion", C. J. Peters and Son, Boston, 2nd ed. 1894 pj 345, said inertia must be looked upon as probably due to motion.

f) Sir Oliver Lodge (English physicist, 1851-1940) in his book "The Ether of Space" 1909, Chapter on the Theory of Light, published by Harper & Brothers, New York and London, believed that, The substance that carried electromagnetic waves needed two characteristics, Elasticity and Inertia. Elasticity so it can be deformed, and inertia, so it can rebound, or in Lodge's words, so it can "Overshoot the mark".

g) From the wisdom of physics texts of the twentieth century, comes this about "**Force**":

A Force is a push or a pull.

and

Force is the only cause of a tendency to bring about a change in the velocity, or direction of motion, or shape of an object.

Waves in the Cosmos

There is a view of the universe that does not emphasize "substance" or the "energy" that drives that substance into action. The view features change, as "**The Way of Nature**", or to use the Chinese term, "**The Tao**". In this view, change is a continuous, ceaseless flow,

The cycle of change is "The way".

To be a Taoist is to recognize nature's cycles and work, with them use them to achieve goals. **Cyclic change is dynamic and occurs between any polar opposites**. The Yin and the Yang, are the most basic polar opposites in Chinese philosophy. The character of Yin is being reactive, to resist change, tend to rest, to procrastinate, to take the easy way. The character of yang is that of movement, action, to be energetic, to promote change.

Waves in the cosmic ocean pulse between the opposing forces, forces of aggressive change (the yang quality forces), and the forces of anti-change of relaxation (the yin quality forces).

Traveling waves wash across the vast expanses of interstellar space, whilst other waves pulse in clumps being the more solid bits of the cosmos.

In this chapter let us explore the nature of simple wave activity. This explanation of wave activity will not be comprehensive but will describe a few of the main ideas concerning waves, such as:

a) What are waves made of?
b) What is a Simple harmonic wave?
c) The three characteristics all waves have.
d) How to represent a wave on a graph.
e) What keeps a wave going?

The waves dealt with here are waves in "Simple Harmonic Motion", they are all of a single frequency, and of a single maximum amplitude.

In This Theory

a) Waves consist of the motion (displacement) of Primordial substance driven by energy. Many such waves are undetectable by man's five senses.

b) The Simple Harmonic Wave is the repeated motion of a body or particle between two extremes, back and forth across an equilibrium position. A swinging pendulum exhibits this sort of motion.

On a graph the equilibrium, or cross over position (the neutral, or zero displacement position) is represented by the horizontal axis. The displaced substance behaves like a steel spring that has been compressed. The farther from neutral position the substance or particle is displaced, the greater the restoring force will become to return it to neutral.

There are two halves to a "period" of simple harmonic motion. One is when the particle moves upward away from the zero displacement position and then back to it. The other half is when the particle continues it's downward motion (crossing the horizontal axis on the graph in the process) moving away from the neutral displacement position, and then back up to neutral displacement. This activity is repeated over and over. The displacement line on the graph mimics this repeated up and down motion.

The motion above and then the motion below the horizontal axis, forms the two halves of a complete "period", "Cycle", or "Wavelength". The Frequency of a simple harmonic wave is the period of time it takes to complete one wavelength. The frequency is stated in cycles per second or alternately as "Hertz's" per second.

c) <u>The three characteristics of a simple harmonic wave are:</u>

Displacement, Acceleration, and Velocity.

d) On a graph of a simple harmonic wave, the vertical axis represents the amplitude of any of the three characteristics. The Horizontal axis when dealing with a traveling wave represents elapsed time. When dealing with a standing wave the horizontal axis represents an instantaneous condition of the wave over a distance.

e) What keeps a wave going after displacement reaches zero?

The answer is "Velocity" of displacement. Velocity is maximum when displacement reaches zero because it is $1/4^{th}$ wavelength out of phase with displacement. Or in the language of physics the velocity lags displacement (it reaches it's maximum $1/4^{th}$ cycle after displacement does.

The Three Characteristics of a Traveling Wave

Displacement, Acceleration, and Velocity

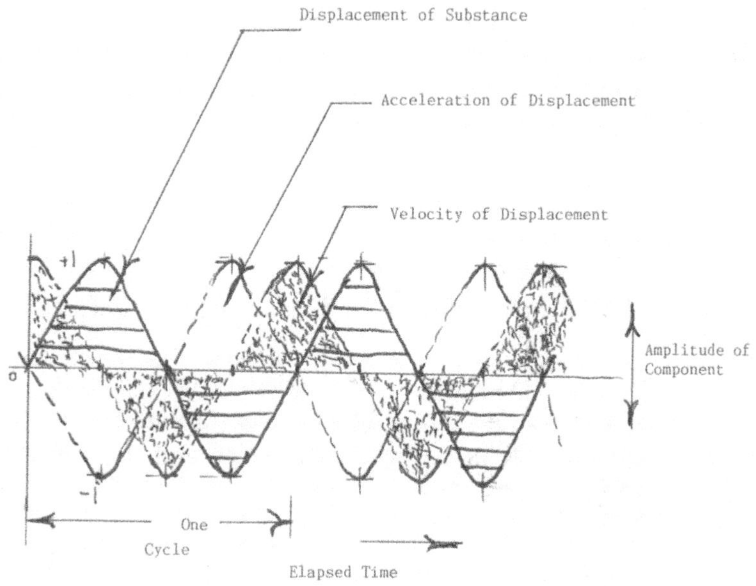

Fig. #18

Location of the three wave characteristics
in a Matrix diagram,

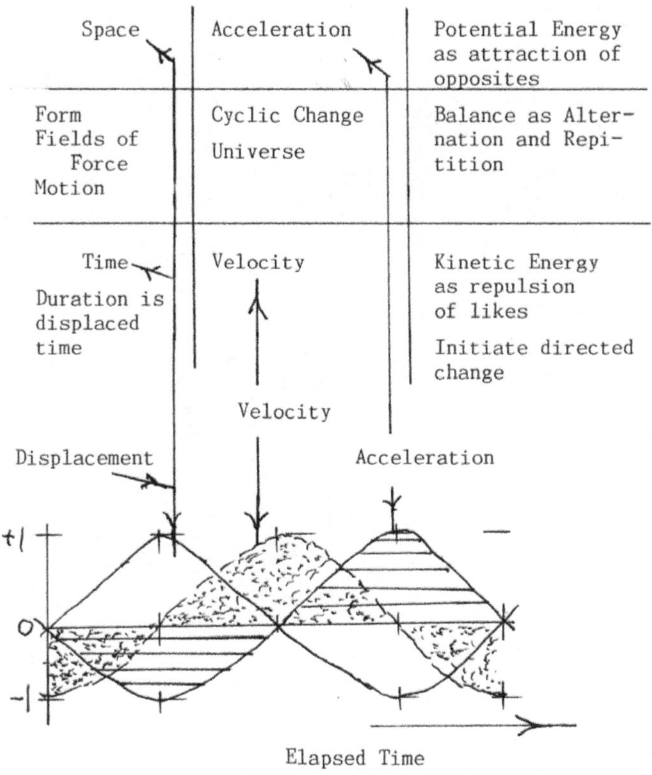

Fig. #19

131

Graph of the Electromagnetic Wave

This is the standard text book presentation which does not show the velocity characteristic

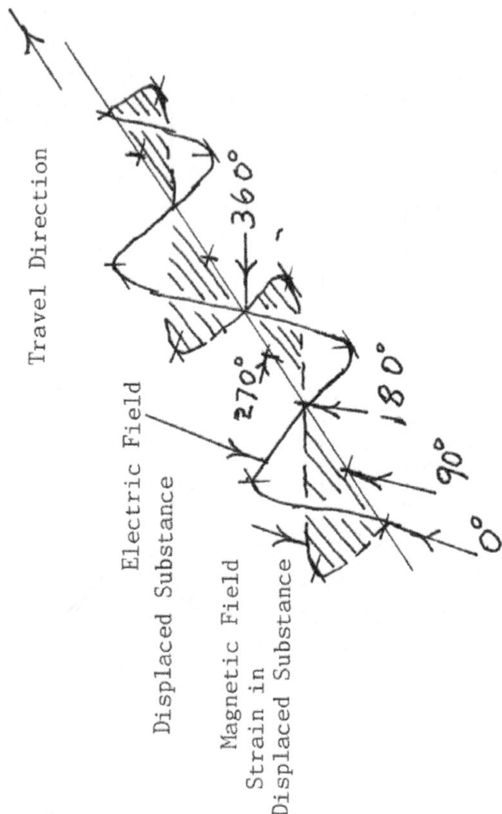

Fig. #20

Simulated 3D graph of wave.

Graph of the Electromagnetic Wave

This is not the standard text book presentation
of an electromagnetic wave, as it includes
the velocity characteristic of the wave.

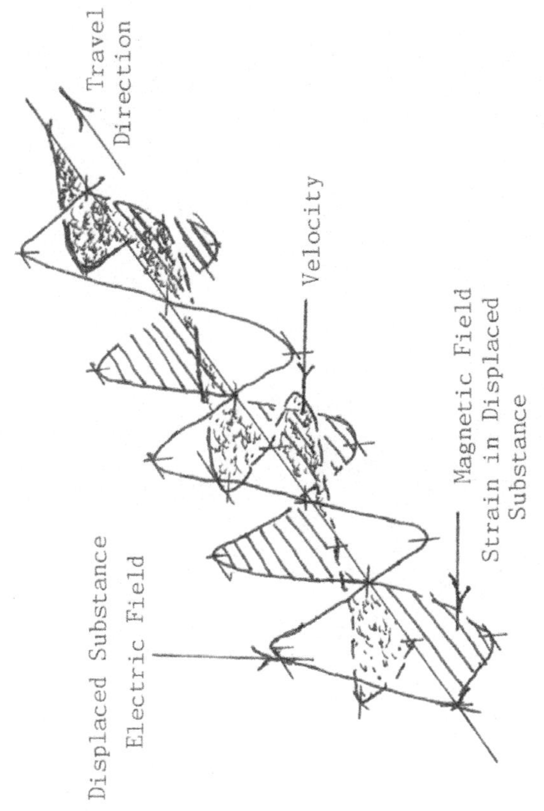

Fig. #21

As you can see velocity rises and falls <u>within the wave</u>. This
is a simulated 3D graph of wave.

Simple Harmonic Wave Motion

This is a flat version of the graph in Fig. #21 Simple harmonic wave motion is depicted. This graph is for a single frequency and includes velocity.

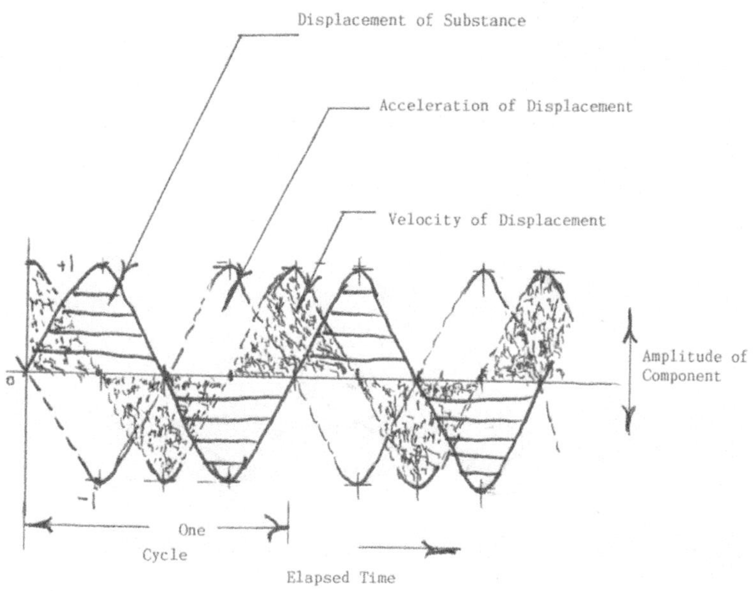

Fig. #22

The Velocity of Electromagnetic Waves

<u>Standard physics texts</u> state the speed of light is constant in a vacuum. In transparent materials the speed of light is slower. A table of the indices of refraction of different materials will let you calculate the speed of light in each of the materials, as follows:

speed of light = <u>**Speed of light in a vacuum**</u>
in material   Index of refraction

There are two vary interesting phenomena related to the travel of light waves or other electromagnetic waves.

1) When passing into an optically dense material that slows the waves the curious thing is when they re-emerge out the other side, .they are right back up to speed again, it is as if the substance acts like a spring, on entering the substance this imaginary spring is compressed and when the wave exits the substance, the spring gives the waves a kick of energy and they are back up to speed again as if being in the material had never slowed them.

2) When light waves pass across the boundary separating media of different optical densities, the frequency remains the same but the wavelength changes.

Standard physics texts for purposes of rough calculations give the velocity of light **in a vacuum** as constant at: 300,000 kilometers per second (non-metrically it is, 186,000 miles per second.) For more accurate work the value experimentally arrived at is 299,972,458 kilometers per second.

Standard physics texts say an electromagnetic wave of displacement, consists of an electric field and a magnetic field that is 1/4th wavelength out of phase with the electric field. These two fields are oriented at right angles (90 degrees) to one another. These fields move in a side to side motion (a transverse motion) as the wave moves down it's path of forward advancement.

Fig. #20 is the standard physics text version of this simulated three dimensional graph.
Fig. #21 adds the velocity component to the simulated three dimensional graph.
Fig. #22 is a standard flat two dimensional graph of the wave in Fig. #21.
The waves in these three graphs are "traveling waves" they indicate the amplitudes of each of the three wave characteristics during each instant of time, over a series of instants. The horizontal axis of these traveling wave graphs is elapsed time, the vertical axis is amplitude.

The Speed of Waves In This Theory
is not constant

In this theory the speed of light, like that of other waves in Primordial Substance is not fixed. The speed of waves is set by the "strain level" in the space the wave is passing through. **More strain, faster travel speed.** The apparent constant speed of light in vacuum only means the strain level varies, very little under the usual conditions it is measured at.

The Standing Waves

Firstly a standing wave is a traveling wave that is being repeatedly reflected between two reflectors. The wave reflects back and forth between the reflectors reversing phase as it reflects. For an explanation of phase see pages 189-191. Alternately a standing wave can be produced by colliding traveling waves of suitable phase.

Standing waves have fixed zones that alternate along their length. A zone with maximum activity is termed an "Anti-node" this is followed by a zone of minimum activity, termed a "Node".

Each of the three wave characteristics has it's own, **Node and Anti-Node zones**. Those for Displacement and Acceleration pulse simultaneously and therefore occur at the same positions. Velocity nodes and anti-nodes appear $1/4^{th}$ wavelength out-of-phase compared to those of displacement and acceleration.
This means that where displacement and acceleration exist as an anti-node zone, velocity will have a node zone. Where velocity exists as an anti-node zone, displacement and acceleration will exist as a node zone.

What occurs at a node zone in a standing wave?

There is very little fluctuation of displacement amplitude at a node zone. In a sound wave in the atmosphere, a displacement node would be at atmospheric pressure. **Fig. #24** shows a standing sound wave at peak displacement during a half cycle, in the bore of a clarinet.

The standing wave in the bore of a clarinet

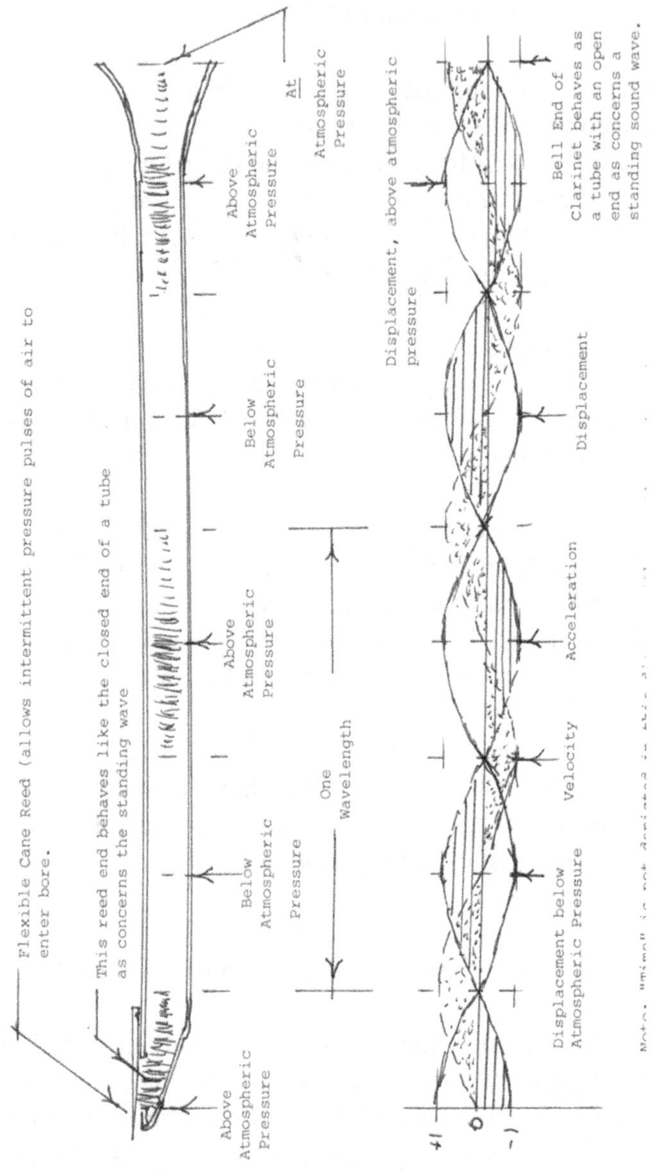

Fig. #24

What occurs at displacement anti-node zones along a standing wave?

The amplitude fluctuates on either side of a neutral value of a displacement anti-node zone, in a manor of speaking the amplitude flip-flops in a smooth flow between a maximum value above the neutral value (the graph's horizontal axis), to a maximum below the neutral value. In a sound wave, displacement amplitudes above the horizontal axis on a graph are above atmospheric pressure, and displacement amplitudes below the horizontal axis on the graph are below atmospheric pressure.

The displacement anti-node zone activity can be likened to soldiers marching in place, their feet go up and down but the soldier neither moves forward nor backward, in a sound wave, it is the pressure that goes up and down. In a torsion wave in primordial substance, it is the degrees of torsion rotation that vary and reverse direction.

Each graph in **Fig. #25** shows a frozen image of the three wave characteristic amplitudes at maximum displacement, for a sequence of two half cycles of a standing wave.

Note In a standing wave the horizontal axis does not indicate elapsed time as it does in a traveling wave. In a standing wave the horizontal axis indicates physical "Distance" along the standing wave. Sound wave node and anti-node zones can be shown to exist experimentally in a transparent tube using cork dust, which tends to collect in the displacement node zones.

Fig. #26 Shows the location of nodes and anti nodes on a graph of a standing wave, along with their relation to "end conditions". The clarinet (and all wind instruments) have **"end conditions"**.

Most wind instruments like horns, clarinets and saxophones as far as standing waves are concerned, have one end considered closed (the reed or mouthpiece end) and the other end (the bell end) as an open end. The Flute and Piccolo are different in that both ends as far as standing waves are concerned are considered, open ends.
String instruments, have two "Fixed ends to the strings so they can be tensioned.

If you wish to learn more about sound waves generated in musical instruments the following book is excellent on the subject,

"The Physics of Music" published by W.H. Freeman and Company c 1978 and various earlier dates. ISBN 0-7167-0096-4 and ISBN 0-7167-0095-6, pbk. This book has reading from Scientific American.

Graphs of Maximum displacement
amplitude during each half cycle.

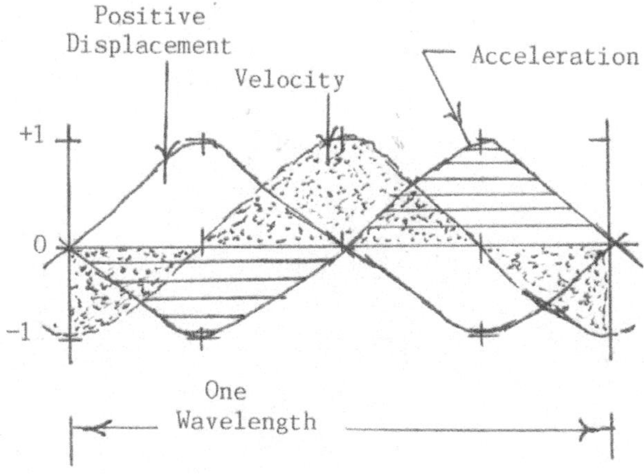

Time wise this is the first half cycle

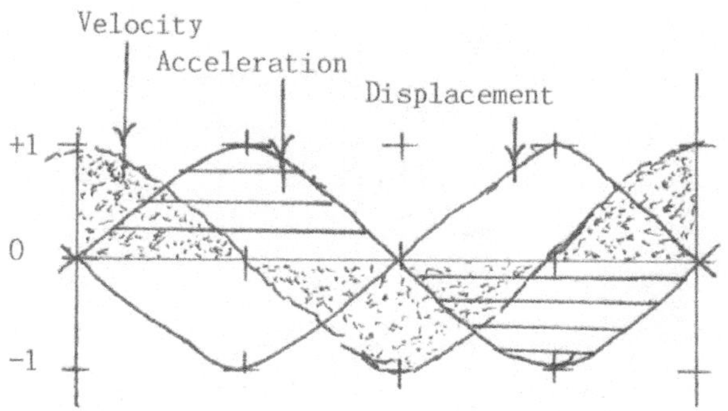

Time wise this is the next half cycle

Fig. #25

Location of Nodes and Anti-Nodes along the length of a standing Wave.

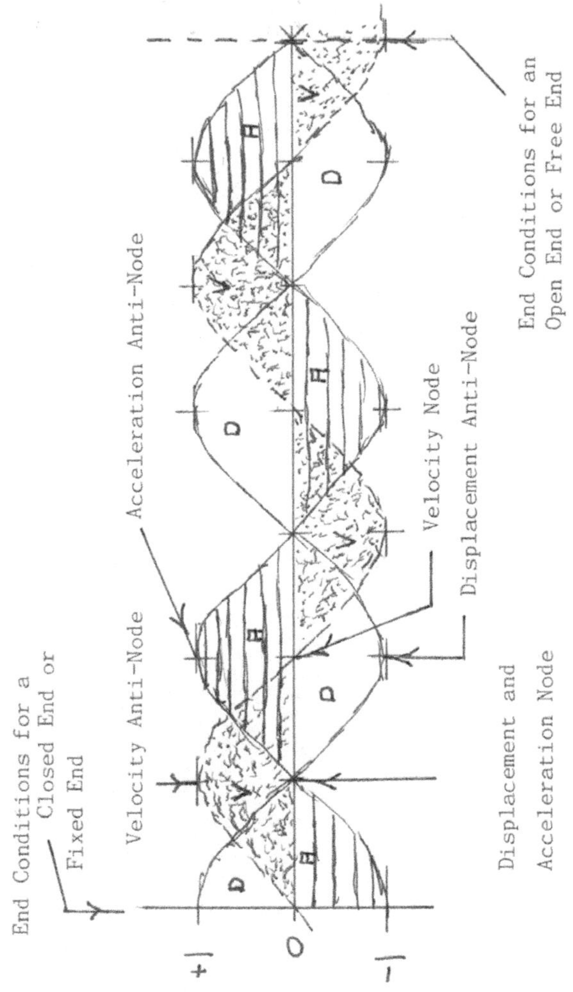

Fig. #26

Accompanying notes for wave graphs

Note: <u>Displacement and Acceleration nodes</u> along the standing wave <u>are velocity anti-node</u> zones. Displacement and Acceleration anti-nodes along a standing wave are Velocity nodes.

Displacement nodes are one-half wavelength apart that is 180 degrees.

Velocity is $1/4^{th}$ wavelength out-of-phase with Displacement, that would be 90 degrees out-of-phase. Velocity is said to "Lag" Displacement.

What the graphs of
Fig. #20 & Fig. #21 Represent

There are some who believe these diagrams are pictures of actual electromagnetic waves. These illustrations are graphs of the three wave characteristics over an elapse of time (for traveling waves), or over a distance (in a standing wave).

The graphs of Figs. #20 & #21, to better visualize what is occurring more clearly, show a sort of simulated three dimensional view of the graph termed an "Isometric drawing".

Table S
End Conditions at Displacement Nodes and Anti-Nodes

Conditions that produce reflection at a displacement anti-node	Conditions that produce reflection at a displacement node
For the free end of a rod rope or string	For the fixed end of a rod rope or tensioned string
For Sound Waves	
The Closed end of an air column	The Open end of an air column
When sound waves reflect off off an extended reflector like the surface of a lake.	
A wave traveling in a solid or liquid and reflecting off of a gaseous reflector	A wave traveling in a gas and reflecting off of a solid or liquid reflector

Accompanying notes for Table S

Interfaces at sufficiently different strain levels within primordial substance apparently can act as reflecting surfaces. This appears to be the situation with the electron standing wave. The outer perimeter of the spherical electron standing wave of this theory has it's surface facing space, able to lower strain in that outer reflecting zone. The space well beyond the electron perimeter zone is of relatively higher strain.

As noted in Fig. #27, The outer perimeter of the electron wave is a displacement node (and simultaneously a velocity anti-node).

If strain conditions at the perimeter could be reversed, what would be the result?

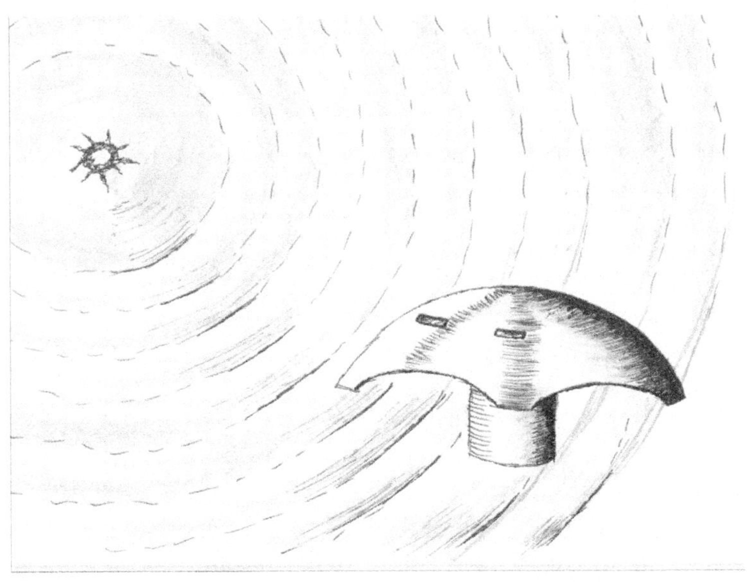

Riding The Waves Of The Cosmos

The Spherical Standing Wave

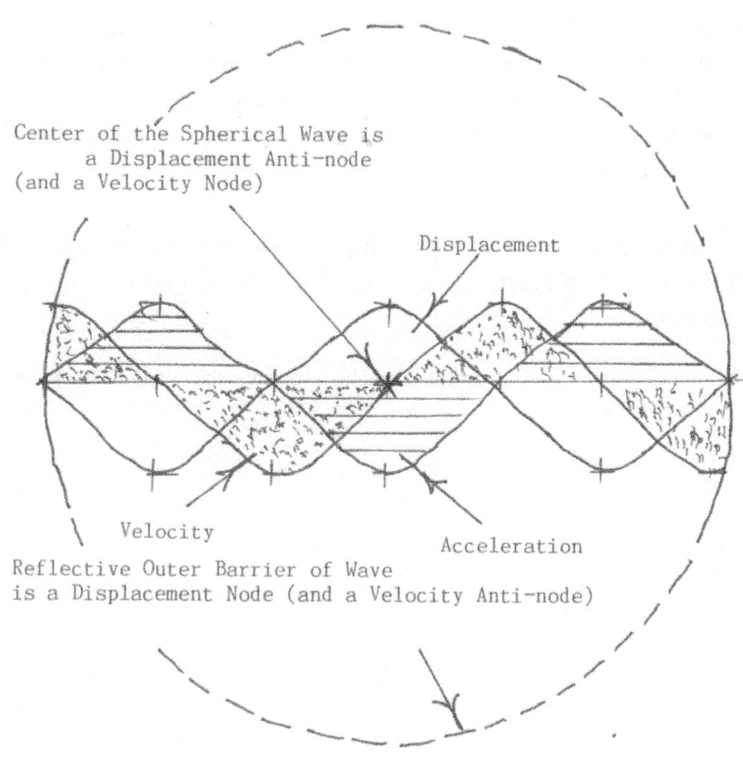

Fig. #27

In this theory, this "Matter Wave" is a Torsion Wave. This means the displacement is a twist motion, <u>at the peak</u> amplitude <u>of displacement</u>, the direction of <u>twist reverses direction</u>.

Spherical Standing Waves

Not much has appeared in standard physics text books about spherical standing waves or spherical standing wave structures. However in 2010 spherical standing waves seem to be coming into their own with a few theorists.

The reason I mention spherical standing waves here is, they appear to be the basis of matter, and so their nature is of great importance to man. One day physics texts may feature a whole chapter on Spherical standing waves and their structure, specialists will write complete books on them.

So what is a Spherical Standing Wave?

I believe a spherical standing wave is a spherical traveling wave of the needed frequency reflected back on itself under the needed conditions. The reflector needs to be at a specific distance from the wave's central origin to achieve resonance.

See Fig. #27 Illustrating **a** spherical standing wave in a state of resonance. One reflector of this wave is the center of the sphere, the point of the wave's origin. The other reflector is the spherical perimeter of the standing wave. All spherical standing "matter" waves have the following arrangement of nodes and anti-nodes.

The central point of origin is a displacement anti-node.

The spherical reflecting perimeter is a displacement node.

A notable characteristic of this condition is the level of strain in space due to displacement. I believe the strain at the outer perimeter reflector displacement node is at zero strain or less.

Physicists term this sort of standing wave structure "Mass", or "Matter". The three sub-atomic particles, the electron, proton, and neutron appear to be the smallest of these "matter" standing wave structures.

The <u>chemical elements are made up of</u> various combinations of <u>electrons and positrons (positive electrons)</u>.

Matter brings about a lowering of the strain level in the surrounding space apparently this lowering of strain is due to the presence of the displacement node at the resonating wave's perimeter.

A larger lump of matter causes a greater lowering of the strain level. I believe this is the source of the gravity force. No artificial means of generating a sustained gravity field has yet been discovered.

Space does not need to be curved as in some theories to explain the gravity force. If we imagine lower strain levels as valleys amongst hills of higher strain levels, it is easy to visualize how a lump of matter "gravitates" down from a higher strain level toward a valley of lower strain level. Our sun has a massive gravity field (a very great lowering of the strain level surrounding it.) The planets would roll into the sun's gravity valley, only their orbital velocity keeps them from being pulled into the sun.

Unanswered Questions

There are still many questions unanswered even if the speculations of the theory in this book prove correct. Questions such as:

1) What are the conditions that trigger reflection of the electron's displacement wave at it's outer perimeter (it's displacement node)?

2) What controls the velocity of the displacement of an electron wave? Is it the strain level of the space the wave travels through?

The apparent average velocity of light waves in interstellar space may be due to "strain" from the presence of a huge "universe wave" that takes billions of years to complete even a half cycle, and so, to us, the strain level of this huge wave appears constant over our life time. In this theory, a lower strain level means a slower wave velocity, higher strain level, means a higher wave velocity.

3) What sort of instrument will detect the strain level in space? Could such an instrument also detect variances in the strain level inside an electron wave structure?

4) The Wavelength of known electromagnetic waves ranges from the relatively long wavelengths of electric waves (like those from the electric power company at 60 cycles per second) which are many kilometers long. We then move to shorter and shorter waves. meter length radio waves, infra-red heat waves in the micro-meter range (10 to the minus six power meters), visible light waves with wave lengths 1/100th the width of a human hair (740 t0 440 nano-meters, that would be ten to the minus ninth power meters) to the even shorter invisible ultra-violet waves (that make fluorescent materials fluoresce, and then the extremely short, atto meter range (ten to the minus eighteenth power) hard gamma waves. The question is:

Is there a limit to how long or how short these waves can be? Perhaps their wavelength range is infinite.

Side Light: Some believe the space of the universe is "Isotropic" that is, that it has the same character in all directions. Others believe space is anisotropic, that is not the same in all directions.

If a "Torsion Wave" the size of the universe, a "Universe Wave" exists in space, space will not be isotropic it will be Anisotropic, and have directionality to it, the direction of twist of a huge torsion wave might well give it that directionality. The space of this theory is "strained" due to being displaced. Interstellar space strain levels may not vary much but will be present. They may have already been found by their effect on light waves from a very, very distant special type of star which emits polarized light, the plane of it's polarization is predictable yet when the light reaches earth it's plane of polarization has been rotated a detectible amount. If you wish to read more on this subject see the following works:

a) Borge Nodland's internet page, Anisotropy in Electromagnetic Interactions.

b) Indication of anisotropy in electromagnetic propagation over cosmological distance, by B. Nodland and J.P. Ralston, Physical Review Letters, 78:3043:3046, 1997.

c) Primordial Torsion Fields as an Explanation of Anisotropy in Cosmological Electromagnetic Propagation. Modern Physics Letters A, 12:3003 by

A. Dobado, and A.L. Maroto, 1997. see http://xxxlanl.gov/abs/astro-ph/9706044.

d) A Glimpse of Cosmic Anisotropy by Borge Nodland

e) The Relevance of Directions in the Cosmos by Borge Nodland, approximately 23 pages. The theory in this book sees the waves in the ocean of the universe as "Torsion Waves"

The Reasons" Torsion Waves" were chosen for the waves in this theory

1) In a universe where the substance, "Primordial Substance" is uncompressible and not rarefiable, only rotary motion within it is possible.
<u>Non-torsion motion is possible</u> when a whole spherical standing torsion wave changes location in space. Examples of this are the everyday motion of spherical standing wave objects.

2) Visualize the universe as a huge sphere. In this theory where space has two halves one moveable the other not, and time is similarly seen, The moveable half cannot drift off to one side, outside of where the unmovable half is not. The moveable half of substance when it moves bodily can only move as a torsion motion since it is a given that it cannot exist alone without it's complementary opposite other half present.

3) This vision of the substance of the universe allows continuous motion to take place with minimum upset, the space an object occupies is not transported across the universe, only it's wave activity is transferred from space to space, to space, like the motion of waves of an earthly ocean are transferred through the water.

Electromagnetic waves in this theory are torsion waves. <u>The electric field</u> of these waves is, substance moving at a displacing velocity.

<u>The magnetic field</u> of the wave is "strain" in the wave substance brought on by it's deformation, it's displacement. Acceleration is a change of velocity of displacement of substance. In defense of this speculation of Electromagnetic waves being torsion waves, I would put forward that a torsion wave appears not all that different from a "Transverse Wave". Yes, a torsion wave has a torsion axis and a directionality to its twist that keeps reversing, but it pulses like a transverse wave.

<u>A Caution to students:</u>

Standard physics texts say, electromagnetic waves are "Transverse Waves".

The mathematical handling of different quantities

Mathematically there are two general classes of quantities, scalar and vector. Scalar quantities have only magnitude and sense (sense here being their positiveness or negativeness).

Vector quantities also have direction in addition to magnitude and sense (positiveness or negativeness).

Scalar quantities are added together arithmetically. Vector quantities are added together geometrically.

Examples of **Scalar quantities** are: Volume, Time, Work, Energy, Speed (linear and angular), Power, Moment of Inertia, and Mass.

Examples of **Vector quantities** are: <u>displacement</u> (linear and angular), <u>velocity</u> (linear and angular), <u>acceleration</u> (linear and angular), Momentum (linear and angular), Force, and Torque.
Note **all three characteristics of waves, are vector quantities.**

A Torsion Wave in a Steel Bar

One end of the rod is clamped, the other is free. The arrows indicate the alternating directions of rotational displacement, along the standing torsion wave's axis, as it resonates
Nodes = low strain in substance.
Anti-Nodes = high strain in substance.

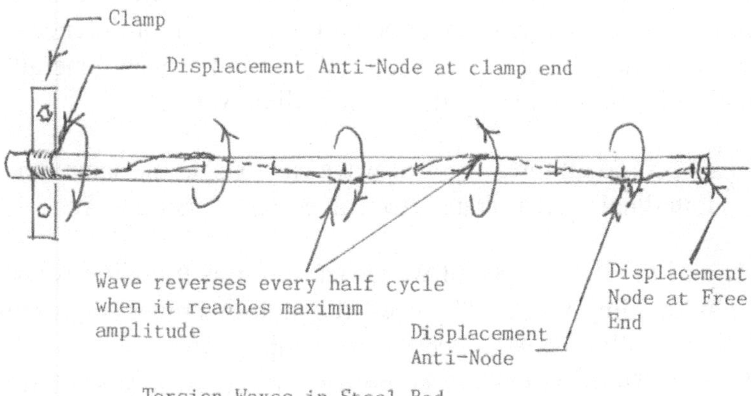

Fig. #28

Indicating the direction of "Twist" of a Torsion Wave
There is an existing practice, a convention in labeling of torque, or rotation as either positive or negative when torque

is graphed, is: When looking at the end of the rotation axis or shaft, if the rotation or twist is clockwise, it is stated as positive torque or rotation, and if the twist is counter-clockwise the torque or rotation is stated as being negative.

The Handling Of Waves

Lenses "bend" waves, mirrors "reflect" waves, and prisms do both. Electromagnetic waves are slowed on entering a material substance.

The type mechanical vibratory wave present in a material object, has to do with how the bar or rod was excited into wave activity, hitting a bar directly on end sets up longitudinal waves. Clamping one end of a bar and striking the other end sideways sets up transverse waves. Torsion waves can be excited in a rod with one end clamped, and bowing a rosined cloth wrapped around the circumference of the free end. In the same material Torsion waves generally have a higher frequency than longitudinal waves.

In a steel rod all three types of wave can exist at the same time, Longitudinal waves, Transverse waves, and Torsion waves.

Any of the three types of waves can exist as traveling waves or as standing waves. The needed conditions must be present for a traveling wave to become a standing wave.

Longitudinal waves are compression waves, in a steel rod they travel from one end to the other. Sound waves are longitudinal waves. The standing waves of sound inside a woodwind like a clarinet are longitudinal waves.

Transverse waves move the substance of a steel bar from side to side as they travel forward down the bar, a serpentine motion.

Torsion waves, are waves of twist, a partial rotation of the steel rod, alternately, twisting a zone of the rod clockwise, then counter-clockwise.

Distribution of Wave Energy

There is a distribution of energy in a chaotic universe a distribution of energy in an orderly universe as wave motion. Wave motion of Substance has both the "Energia" and "Dynamis" the Greek Philosophers spoke of. Today these two complementary states of energy are termed Kinetic Energy and Potential Energy.

Kinetic energy exists in the velocity of the wave substance. Potential energy exists in the wave substance as strain forces due to acceleration.

Strain in substance exists when the substance is displaced it is an elastic force. When a wave is being displaced it's substance behaves like a coil spring being compressed, and when the stress forces due to velocity are reduced or removed, the spring rebounds returning the substance to it's un-displaced condition.

The Cause of Inertia

If someone should ask what is the cause of the property of "Inertia" of a mass? I would say it is due to two characteristics of spherical standing waves. I limit this to spherical standing waves because traveling waves etc. do not seem to exhibit this property.

The "Innate Force" of Newton, I see as due to **The "Acceleration" of wave substance**. The innate force exists in displaced substance as "strain forces, a stored elastic rebound forces.

The ability to "Overshoot the mark" as Sir Oliver Lodge put the requirement I see as what today is termed **"Inertia"**. I believe this "Inertia" is actually **The "Velocity" of the wave substance**

The Smallest bit of "Matter"
The Electron, is it a particle or a wave?

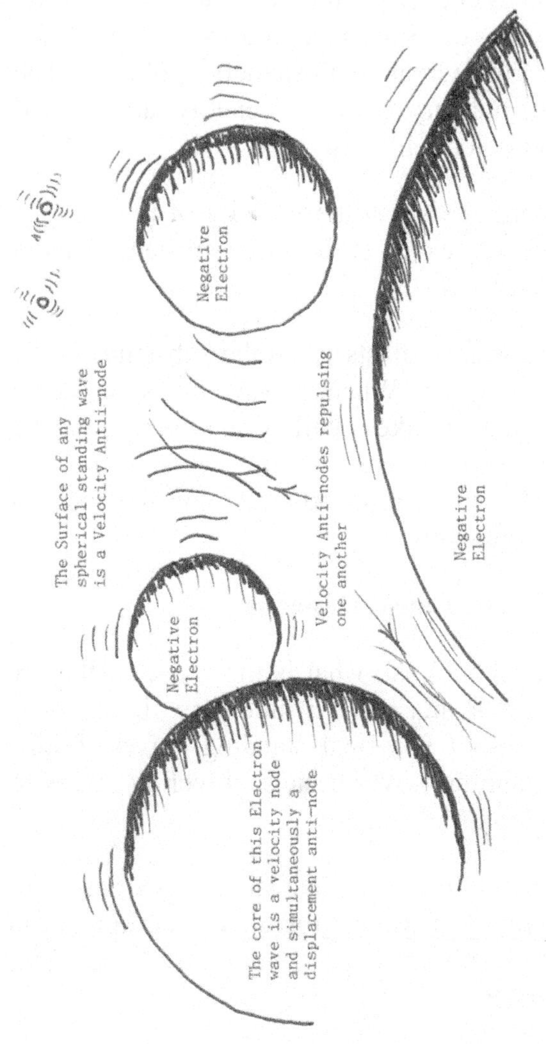

Fig. #29

The Nature of "Matter"

Matter in the theory in this book, consists of spherical standing waves. The most basic forms of matter appear as three subatomic particles, the electron, proton, and neutron. These three are:

The Building Blocks of Matter

According to Canadian theorist **Gabriel LaFreniere**, (who's work, "Matter is made of waves" has appeared on the internet), the electron is the primary particle, and the proton and neutron are actually composed of electrons.

Mister LaFreniere speculates a **neutron** consists of six electrons centered on the faces of a cube, their interacting fields canceling out their negative charge, and that a **proton** is composed of six electrons arranged like those of a neutron but with the addition of a **positron** (a positive electron) located in the center of the cube, giving the proton a resultant positive charge.

Also speculated is that the positron is actually an electron with it's standing wave structure is 1/4th wavelength out of phase with that of negative charged electrons. He goes on to say he believes all electrons in the universe are "in-phase" with one another or nearly so, and that any that are not, soon "fall into line" with the majority.

Where is the extra mass?

Now you may be thinking, the masses of the seven electrons in this theoretical proton do not equal the estimated weight of one actual proton, which equals the weight of 1,837 electrons. So where are the other 1,830 electron masses hiding in this theoretical proton. Now you may think I am a bit credulous in accepting what follows but I believe it is so,

I believe Mister La Freniere is correct. Consider that the electron is wave activity of substance, the positron is wave activity of substance, the mass of a whole proton therefore consists of wave activity of substance. Logically then the missing mass is in the wave activity interactions of the whole proton, and more specifically it is in the **Gluonic wave field** holding the proton's electrons together. A physicist might term this Gluonic field, **"Binding energy". The missing mass is** Binding the seven electrons together and has a far greater mass effect than seven separate individual electron wave fields would have.

The Atoms of the Chemical Elements

The sub-atomic particles in various arrangements make up "The Chemical Elements". There are some 92 "natural elements". Hydrogen is #1. Uranium is #92.

Predicting an element's Properties

How do scientists know what a chemical element's properties are likely to be that has not yet been discovered?

Once atomic numbers were determined for a few of the chemical elements and their properties became known, **Dimitri Ivanovich Mendeleyev** (Russian Chemist, 1834-1907) who did post graduate work in Germany, eventually arranged the known chemical elements in a table of ascending atomic numbers. In 1869 Mendeleyev produced the first successful **periodic table of chemical elements** and used it to predict three then unknown elements including their atomic numbers, densities, melting points and what the chemical composition of their oxides would be. Those elements today are termed element #5 Boron, #13 Aluminium, and #14 Silicon. Since then many chemists and physicists and other persons, have improved and added to the "Periodic Table of Elements". It is from the modern version of the periodic table that element #118 is predicted to be a radioactive noble gas. Will #118 be stable?

Atomic Number

Atomic Numbers were once termed atomic weights, Atomic numbers were not a weight in grams at all, but were comparisons at the time with oxygen. Other element numbers were the result of this comparison, which worked out wonderfully as Hydrogen the lightest element by this method became number one.

Today the atomic number assigned to a chemical element is understood as **the number of protons in the nucleus of an atom** of an element. If there are greater or lesser numbers of protons than a certain element's atomic number states, it means you have a different element. As a matter of interest, A single Oxygen atom (atomic number 16) has a real weight in grams of:

0.000,000,000,000,000,000,000,026,39 grams.

Elements up to #83 are atomically stable, meaning they are not radio-active, with the exception of two. **Technetium #43** (German chemists Otto Berg, Walter Noddack, and Ida Tacke may have discovered this element in 1925 naming it Masurium). In nature Technetium #43 is rare, but Technetium is produced in Tonne quantities as a result of refining spent fuel rods from atomic reactors. Technetium's most stable form is Tc 97 **** it has a half life of 2.6 times ten to the sixth power years, it's least stable isotope Tc 99m has a half life of 6.05 hours and is the isotope used for medical purposes.
Promethium #61 - All forms of Promethium are very radio-active, it's most stable form has a half life of roughly 4 million years, it's least stable isotope has a half life of six hours.

The Word "Isotope"

The word isotope means, "same place" and refers to occupying the same place in the periodic table of chemical elements. So isotopes of hydrogen all occupy hydrogen's square in the periodic table.

Hydrogen and It's Isotopes

Hydrogen has three variations of it's structure termed "isotopes", an ordinary hydrogen atom's nucleus is termed "Protium" and has **one proton and no neutrons**. Over 99 percent of all hydrogen on earth is of this type, it is stable and not radioactive. The other two Hydrogen isotopes are rather famous, Deuterium (popularly termed heavy hydrogen), and Tritium. Deuterium has a nucleus with one proton and one neutron, it is stable and not radioactive, about one Deuterium atom exists for each 6,400 ordinary hydrogen atoms on earth. The third isotope of hydrogen is named Tritium, it's nucleus has one proton and two neutrons. Tritium is radioactive and has a half life of 12.5 years there is only the barest trace of natural tritium on earth.

The Half Life

Half-life - after each **half-life** period only 1/2 of the element isotope that existed at the beginning of the period will be there at the end. This means if the half life is one year, after one year only half of the original isotope will be present the remainder will have decayed into another isotope or another element. After two years, half of the half will be left of the original isotope, after three, years half of the half of the half of the original amount will be left and so on.

Atomic Mass

**** You might be wondering, What does this number 97 as concerns Technetium represent?
The 97 represents Technetium's Atomic Mass. Atomic mass is not the weight of an atom of an element in grams,

Atomic Mass represents the number of protons plus the number of neutrons

in the nucleus of an element. Using Technetium as an example: Technetium 97, has

97 minus 43 protons = 54 neutrons

in it's nucleus, whereas atomic mass Tc 99, has 99 minus 43 protons = 56 neutrons in it's nucleus. It can also be seen from this example that adding more neutrons affects stability of the atom.

Once we advance past element #83 (non-radio-active)Bismuth to element #84 Polonium a highly radio active element, and probably the most toxic of all the elements to man, and proceed to higher atomic numbered elements, we find they are all radio-active. Element #86 Radon is a gas at room temperature and radioactive, the only radio-active gas (so far discovered) We then proceed up to man made element #118 which is predicted to be a radio-active noble gas similar to radon.

Slowly elements with numbers greater than element #92 Uranium are being synthesized in nuclear reactors or by using accelerators, Elements up to element #103 Lawrencium have been named so far (year 2010), other elements up to #116 have apparently been "made", with the exception of numbers #113, #115, #117, and #118. There may of coarse be more elements above #118, perhaps one day they also will be "made". Only very tiny quantities of some of these new heavy elements have been made, barely detectible amounts in some instances, and some are very unstable, #112 half life is less than a millisecond. "Matter" has other "gross properties" discernable directly by human senses such as:

a) Matter .occupies a **volume** of space.

b) Matter **can be touched**.

c) Some matter has an **odor**, it can be smelled, such as Osmium #76, Osmium is perhaps the densest element of all, it weighs per unit volume more than gold, lead, or uranium. weighing in at 22,590 kg. per cubic meter, (49,820 pounds).

d) Matter is the **source of the acceleration of gravity**, No artificial source of gravity has been found as of 2010 and made public.

e) Matter has **"Weight"**.

f) Matter has **"Inertia"** that is it has a resistance to having it's direction of motion or velocity changed. It takes a force of some kind to do either.

g) Matter has **color**, as perceived by humans. (it reflects certain frequencies of electromagnetic waves out of those striking it, and absorbs or is transparent to others. Examples are: Carbon-black, Sulfur-yellow, Copper-Orange-red, Chlorine gas-greenish, Bromine-a deep red, Nickel-silvery color with a faint yellow tint, etc. For Superman fans, the real element Krypton #36, is perhaps the rarest gas on earth, it is colorless and odorless and exists in the atmosphere.

h) Matter has "State", that is it exists as a solid, liquid, gas, or plasma. A plasma being an ionized gas containing both plus and minus ions, a plasma is a good conductor of electricity and is affected by a magnetic field. Matter is the organized form of the body of the Universe, it is unique. Without matter no "created" order would exist, we would not exist, the "things" of our world would not exist.

Other properties of Matter

Other properties of matter include: elasticity, brittleness, flexibility, density, melting and boiling points, vapor pressure, magnetic and electrical properties, hardness, crystal-structure, some matter has a non crystalline state (amorphousness), chemical, electrical, and atomic properties, etc.

The Location of Chemical Elements

Many pure elements are locked in minerals that need processing to get them out. Some of these minerals are found in only a few places on earth. Helium gas exists in only tiny amounts in earth's atmosphere, and was rare until it was discovered in the gases of some gas wells (about 1895).

The sea which covers almost 3/4 ths of the earth's surface area holds a surprising weight of dissolved chemical elements in it. A veritable treasure house of material is there for the taking, most of the 92 natural chemical elements including, absolutely huge amounts of **Chlorine**, **Sodium** and **Oxygen** and those listed below are to be found in the sea.

In a cubic kilometer* of sea water are to be found:

68,400 metric tons of **Bromine**
1,300 metric tons of **Magnesium**
1,000 metric tons of **Aluminium**
60 metric tons of **Iodine**
10 metric tons **Copper**
200 kilograms of **Uranium**
26 kilograms of **Gold**
2 kilograms of **Silver**

The problem is no one has found a low cost way to separate the water from the solids and gases ** dissolved in it, boiling is to expensive. One day, one day someone will find a way to recover the treasures of the sea at low cost.

The resources of the sea, like those on land, though vast, can also be mined in excess to the determent of the creatures that live in it. Moderation lads, moderation.

Conversion factors

*** One cubic kilometer = 0.02399 cubic miles**
  One cubic mile = 4.168 cubic kilometers
One Kilogram = 1,000 grams = 2.20462 pounds
  One pound = 0.4536 kilograms
1,000 cubic centimeters = 1 liter volume
      = <u>about</u> 1 kilogram of pure water.
One Metric Ton = (Tonne) = 1,000 kilograms
Avoirdupois ounces are used to weigh groceries.
One Avoirdupois Ounce = 28.3495 grams.
Troy ounces are used to weigh precious metals,
One Troy Ounce = 31.1035 grams.
One Gram = 1,000 milligrams

** Except perhaps a device termed "The Evis Water Conditioner" made back in the 1950's by the Evis Manufacturing Company, U.S.A. It caused dissolved gas in water to leave solution simply by flowing through it. I know for a fact that pouring a can of carbonated soda water through it rendered the soda dead flat as fast as it went through the device. The Evis had no moving parts, used no added chemicals, electric current, or strap on magnets. The Evis apparently depended for it's gas removing action on water contact with it's specially treated metal interior surface. As far as I can recall the devices were made in bronze, and cast iron versions, whichever the customer wished, and in pipe sizes from 3/8" id to 6" id. I cannot attest to whatever other claims were made for the device, but it was a wizz at the one described above. I hope it's special metal treatment technology is not lost. The Evis device did not precipitate out dissolved solids.

The Size of Chemical Element Atoms

If chemical element atoms were solid spheres they would have diameters from one to five Angstrom units. One angstrom unit equals ten to the minus eight power centimeters. The largest atoms are those of the light metals Sodium, potassium, calcium. Among the smallest atoms are those of the elements Carbon, Osmium, Nickel and Ruthenium.

The Size of the Nucleus

The Nuclear radii sizes lie between 1.5 times ten to the minus thirteenth power centimeters and ten to the minus twelfth power centimeters.

The Size of the Electron

J. J. Thomson (1856-1940 Sir Joseph John Thomson, English physicist, discoverer of electrons), estimated electron size to be roughly one fifty thousandth the radius of an atom, or a radius of 2.8 times ten to the minus eighth power centimeter. In the theory in this book this electron would be a standing wave structure.

If you would like to learn more about the chemical elements, these two books by John Emsley are very informative.

"Nature's Building Blocks" by John Emsley, Published by Oxford University Press, c 2001, ISBN #0-19-850340-7. Includes brief histories of all the elements and many interesting facts about them.

also

"The Elements" by John Emsley, Published by Oxford University Press, c 1989 and with corrections in 1990, ISBN # 0-19-855238-6 and ISBN # 0-19-855237-8 (pbk). This is what I would term a technical data manual on all the elements.

Speculations on Using an
Artificial Electron

In my previous work "The Man Who Saw Space and Time" I mentioned the possibility of making an "Artificial Electron". Before that is done it needs to be asked "Why is the natural Electron the size it is?", I believe it is so because of the strain level in space, I believe the strain level sets the propagation velocity of the resonant wave, this sets the electron wave length. At the outer perimeter of the spherical electron standing wave is a reflector. The outer perimeter is a displacement node. In the center is a displacement anti-node.

The natural electron spherical resonant wave structure is amazingly impervious to intrusion that would undo resonance. Our "Artificial Electron" may not be quite so resistant to destruction, also to be kept resonating it may need a continual "trickle charge". Still I believe this artificial electron would be very useful.

To construct an artificial electron, we need to choose a wavelength. Perhaps a wavelength of from one half to one meter would be useful (1.64 to 3.2808 feet). Then we need a chamber to act as a containment vessel for our resonating electron wave. It will need to have it's inner surface, or the zone adjacent to it act as a reflecting displacement node. Once we get our artificial electron wave resonating, **it will have a tremendous "mass"** far more than a billion, billion natural electrons.

Gabriel LaFreniere a Canadian theorist in his work "Matter is Made of Waves" proposes that the natural electron is actually a spherical standing wave structure in the substance of the aether.

Milo Wolff an American theorist believes the natural electron spherical standing wave is a wave in the substance of space itself.

The Natural Electron as a Power Source

Common sense tells us there is always a price to pay for anything done in the universe. The natural electron holds what can be seen as potential energy, that is stored energy in it's wave field.

The electron wave energy held as an elastic force (a strain field) in the displaced Primordial Substance could be harnessed as a power source for use by man. This might be done by converting the extremely high frequency and short wavelengths of the natural electron wave into longer wavelengths of lower frequency.

The process of lowering the electron frequency could be likened to that of getting a high frequency tuning fork to cause a lower frequency tuning fork to resonate at it's lower natural frequency.

Solve how to do this with an electron, and the electron energy can be siphoned-off for man's uses to make the world a better place for all living things.

A second method to extract power from natural electrons is by changing their phase slightly and siphoning off the power difference.

See the sections of this book on "The universe as Balance" and "The Gravitational Force" for further comment on the resonating spherical standing electron wave, and it's character.

Fig. #30

The Gravitational Force

Sic itur ad astra   (Latin)

Such is the way to the Stars (English)

The Nature of the "Gravitational Force"

"Gravity" by definition is the "Weight" or "Heaviness" of an object of "Matter". Each piece of matter attracts all other pieces of matter in the universe by means of this "Gravitational Force". The reason we are held to the surface of the earth is due to the gravitational force.

Outstanding characteristics of the Gravitational Force

1) Gravitational force is generated by the spherical standing waves that constitute "Matter", that is, that make-up matter. The form these spherical standing waves take is as subatomic particles, the electron, proton, and neutron, each is "matter" and the atoms they become part of are the chemical elements of our world.

2) It appears that during the life-span of matter the intensity of the gravitational force does not diminish. Only the destruction of matter stops the production of the gravitational force.

3) No matter known, shields against, absorbs, reflects or neutralizes the gravitational force. No matter is immune to gravity.

4) Both electrically positive and negative matter generate an attractive gravitational force. both matter and anti-matter attract one another by gravitational force. No matter has been discovered that emits an anti-gravitational (repulsive) force.

5) Only distance from the point of origin of a gravitational force appears able to diminish it's intensity, or as Isaac Newton put this situation: Any two bodies attract each other with a force that is directly proportional to the product of their masses and inversely proportional to the square of the distance between their centers of mass.

6) The gravitational force apparently extends to the limits of the universe from each source generating it. The rate of propagation (rate of travel) of the gravitational force is unknown, but is thought to be comparable to the speed of light waves.

7) The more matter (mass) that is present, the stronger (the more intense) is the attractive force of the mass. A spaceship positioned between two large masses such as planets will feel the tug of both planets, however it will move in the direction of the one who's attraction it feels most. There will be a small zone where both planets attract the ship equally, The ship can sit in this zone and be un-moved by the planets gravity, however this neutral zone itself moves due to the orbital motion of the planets. See page 53 for more on this.

8) Matter in a gravitational field is subjected to uniformly accelerated motion.

9) Most fields of force have opposite poles, magnetic fields have a north and south pole, Electric fields have plus and minus charged poles. But gravitational fields appear to have only one sort of pole, A pole that attracts all matter, and no opposite pole that repels matter.
See page 175 for more on this.

Questions as yet not answered about
the gravitational force:

1) Exactly how is the gravitational force generated by the spherical standing wave structure, that is by the electron?

2) Does the gravitational force have a frequency and wavelength? It is after all, the result of a wave phenomena.

3) What is the speed of propagation of the gravitational force? When an electron is brought into existence, how fast does it's gravitational field travel outward from it? Is it the same as the speed of electromagnetic waves?

4) Can the gravitational force be generated artificially? Can it's counter force, anti-gravity, be generated artificially?

5) Once generated artificially can the gravitational force be produced as a coherent beam (like laser light) or be focused, blocked, reflected, or absorbed in some way?

If man eventually learns to control and generate the gravitational force I hope he will use it wisely, and not design buildings with a structure so weak they need a reduced gravitational force to stay standing and not collapse.

Finding how to control the gravitational force has become a most discouraging and frustrating exercise. Discouraging because after all the years mankind has existed the secret of gravity control has not been found, if it has, no evidence of it has been made public, and frustrating because here we have a "thing" a "force" that is with us all hours of the day and night, available to be constantly measured, probed, and examined, and still it is not known "how it is made", or it's exact nature.

Cavor and "Cavorite"

Herbert George Wells (1866-1946, English novelist) imagined a trip to the moon in his 1901 tale "The First Men in the Moon". In the tale, a scientist-inventor named Cavor discovered a substance that could block the gravity force much as a wall of brick, blocks light rays, he called the substance "Cavorite". It was compounded in a melted state and when it cooled, and hardened it's gravity blocking property became active.

Cavor built a space ship consisting of an air-tight glass sphere with an access hatch, around the sphere was a protective framework of steel in polyhedral shape, each face of the polyhedron was empty except when Cavorite coated roller blind was drawn across it. With the all blinds closed the ship and it's contents were not attracted by gravity at all, nor was any atmosphere above the ship. To take a trip to the moon he just opened the blinds facing the moon and was attracted toward the moon and woosh off the ship went. After take-off his passenger Bedford commented, It was not like the beginning of a journey; it was like the beginning of a dream. There were no billows of noxious gases as from a rocket, no danger of rocket motors exploding.

During and after World War II, a war in which 12 million people died and some four or five times that number were injured, maimed or starved, a roomer circulated that the German SS military had an antigravity space-ship, there were even witnesses who saw it, and a few photographs of this craft on the ground and in the air. After the war the secret files of the German SS military were examined and wonders, three or four designs of such anti-gravity ships were found with a brief description of each including a side view line drawing of each ship. One such craft was named "Haunebu II". It was said that six or seven of this model in diameters from 26 to 32 meters with a height of 4.25 meters were produced.

The Haunebu II craft purportedly had a crew of 9 persons, and could carry a total of 20 persons including crew, it had a maximum flying time of 55 minutes, a maximum test velocity of 6,000 kilometers per hour (3,728 mph), and the capability of space flight.

It does not take much imagination to realize what sort of "Battle Wagon" such a craft could be turned into. Near the end of the war, **General George Smith Patton** (1885-1945, U.S. army officer and tactician, nicknamed by his troops "old blood and guts"), ordered his armored units to capture the Skoda Works, and SS facilities at Pilsen and Prague, including the factory where the rumored anti-gravity craft were built, and retrieve any craft and documents relating to them and other military technical developments. Patton's troops retrieved two truck loads of documents, no list of what was found has ever been made public. The war diaries of Patton's armored units (Combat Command B, Third Armored Division) at that time cannot be found in the U.S. national archives.** As for General Patton himself, a man who knew what went on in Czechoslovakia at Pilsen and Prague and what was found, a man who was known to speak his mind to the annoyance of some, well, while being driven about on December 9,1945 his driver, swerved Patton's vehicle and hit a heavy military truck head-on, Patton suffered a broken neck, any injuries suffered by the drivers of the vehicles and a passenger Patton was showing about, **Major General Hobart Gray** were not fatal. Patton died in a military hospital in Frankfurt December 21, 1945 and with his death any secrets he knew died with him. The result of all this as concerns these anti-gravity craft is officially **complete silence**, even now in 2010, sixty five years latter.

As for the craft itself "Haunebu II" the alleged SS documents, do not tell anything of it's internal structure or describe it's propulsion motor or furnish technical engineering drawings of the craft or it's motor. It is not unreasonable to believe such drawings would be needed to build such a ship. Were these additional engineering drawings removed by unknown parties?

As for the name of the craft, I could find no town or area of Germany with this name. The name appears meaningless in German, unless one notes the only German word with a spelling close to it, that word is "Haube" it means hat or bonnet in English language, and if one looks at the side view of the Haunebu II line drawing it looks quite like the side view of a hat, or bonnet. At most Hauneburg may have been an unofficial name for the place where "The Hat" was kept.

So where is "The Hat" today?

** see page 114 of "Reich Of The Black Sun" by Joseph P Farrell, c 2004, published by Adventures Unlimited Press, Kempton Illinois, U,S.A. ISBN #1-931882-39-8.

The Gravitational Force as seen in
the theory in this book

The gravitational force as seen by this theory is due to spherical standing wave activity of "Primordial Substance" driven by energy. These waves are torsion waves of displacement.

When Primordial Substance is displaced, strain forces appear in the substance. By analogy these strain forces can be likened to the strains appearing in an elastic substance when it is deformed. These strain forces emitted from primordial substance I believe are the gravitational field.

The strain forces are lowest at an electron's perimeter (surface). When "matter" exists in a lump the lowest strain level is on lump's surface (shown as strain level #1 in Fig. #31 page 176,). On moving away from the planet's surface toward outer space, the strain level increases gradually to that of outer space where the strain level is relatively high (shown as strain level #7 in Fig. #31).

The Cause of the high strain
level in space

The relatively high strain level in outer space away from matter is due to the presence of traveling waves, waves that are not spherical standing waves, electromagnetic radiations or huge "universe waves" traveling through space from distant suns.

There is no "Opposite Pole" for gravity unless one sees **higher strain force as one pole** and **reduced strain force as the opposite pole.**

<h2 align="center">Matter reduces strain level
in space surrounding it</h2>

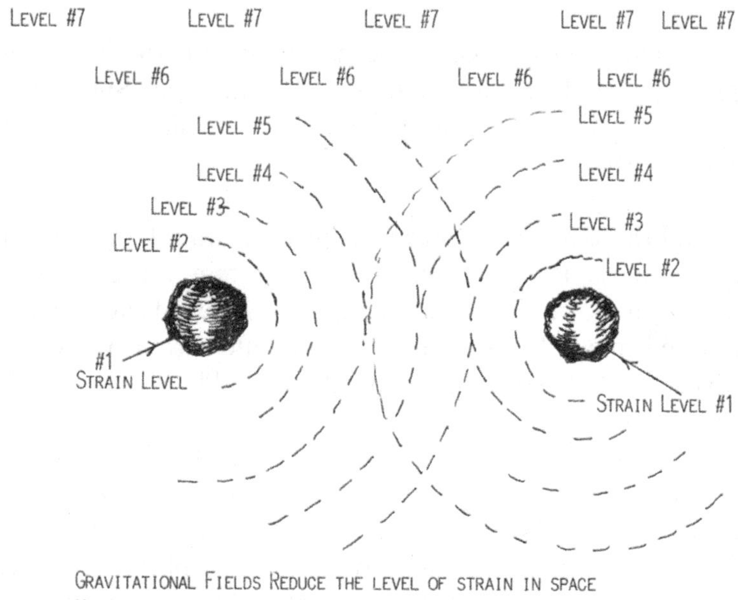

GRAVITATIONAL FIELDS REDUCE THE LEVEL OF STRAIN IN SPACE
HIGHEST STRAIN LEVEL IS #7, LOWEST #1.

Fig. #31

Conclusion about the cause of gravity force

I believe the reduced strain at the perimeter reflector of a spherical electron standing wave is the primary cause of the condition termed gravity.
See Fig. #27 page 146, to see the natural electron spherical standing wave form.

How much power will it take to produce a gravity field? To get some idea of the power needed to produce a gravity field, look at the electron, how much power does it take to produce

176

a natural electron's gravity field? Then scale it up to the size of your artificial election. The natural electron standing wave seems to resonate on and on for eons without adding additional power.

As said an artificial electron will have a huge mass, with an accompanying huge gravity force, but <u>it will not produce an anti gravity effect.</u>
In my previous book "The Man Who Saw Space and Time", I suggested altering the phase of the electron toward that of a positron would produce greater strain in space and anti-gravity. If it's usefulness for space travel proves to be limited, it might be considered for use as a "shield" to protect a space craft.

Floating into the future with Anti-gravity

To bring about an anti-gravity condition, one needs to increase the "strain" in the displaced space surrounding the space vehicle, so it's strain is greater than that in the surrounding space. When this is achieved the space ship will be repelled away from lower strain zones like a bubble rising from the bottom of a pond.

If we follow the reasoning that the perimeter condition (displacement node) of a spherical standing wave causes "gravity", then the opposite condition should produce "anti-gravity". The opposite condition would be a displacement anti-node at the standing wave perimeter.

An inverted electron wave, would have a displacement anti-node at it's perimeter and a displacement node at it's center.

Is Such An Inverted Standing Wave Possible?
Will It Resonate?

I really do not know. See Fig. #32 page #179 for an illustration of an inverted spherical standing "anti-gravity" wave, with it's anti-gravity (high strain) perimeter. For comparison purposes see the natural electron spherical standing wave Fig. #27 page 146.

When standard electron standing waves interact with one another at close range strain levels in the interaction zone vary due to the interacting wave forms, these wave interaction zones form **the "Gluonic Fields"** that hold the electrons in place in an atom and in the atom's nucleus.

These Gluonic fields are limited in extent to their interaction zones and may be attractive (low strain level) or repulsive (high strain level) in nature. This is similar to interference phenomena of light waves only in the instance of these gluonic fields it is direct interacting of these matter waves with one another. The overall effect of these interactions is rather "adhesive", the increased strain gluonic field zones try and repel apart (like charges repel one another) the reduced strain gluonic field zones act like a stretched rubber band pulling the repelling electrons together (opposite charges attract). The outer standing wave electrons of an atom pulse at orbital distances from the nucleus held there by mutual gluonic field interference zone strain levels. Rather like a golf ball settling into a divit.

Install three or four of these inverted artificial anti-gravity producing electrons, plus one gravity producing artificial electron to provide gravity for the crew on your space ship, and "One, Two, Three, Bob's your uncle" you have an anti-gravity propelled ship. It is one of those things that sounds easy, but will probably take considerable doing.

The "Anti-gravity" Spherical Standing Wave

The perimeter of this wave anti-gravity type standing wave is a Velocity Node, and simultaneously a Displacement Anti-node. This is just the opposite node anti-node positions found in the normal electron wave that results in the gravity force. With this reversed condition (the perimeter being a displacement anti-node) an increased strain condition will be present there.

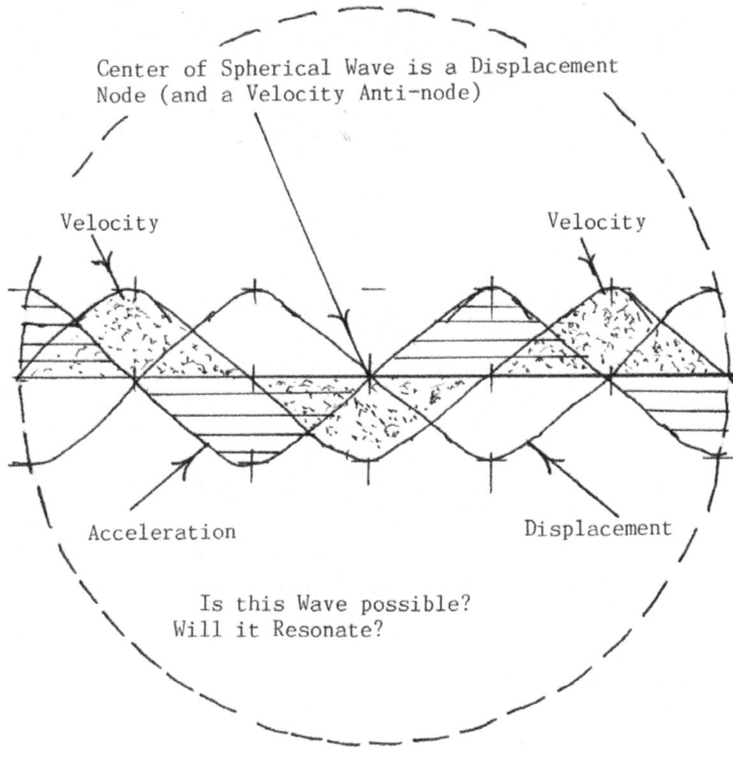

Fig. #32

Travel Time To The Planets and The Moon

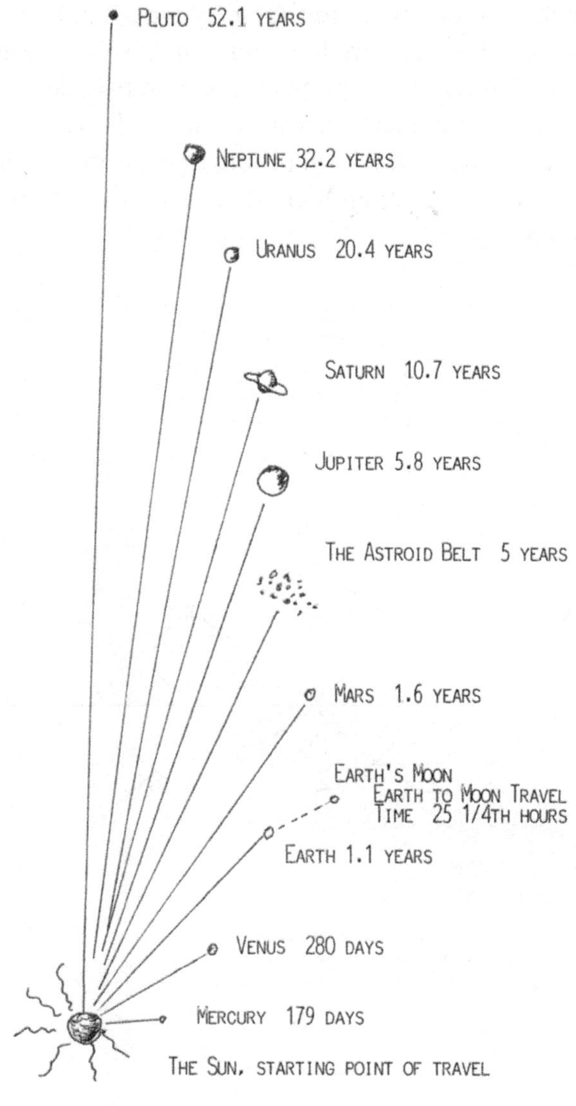

Fig. #33

A tall Ship and a Star to Steer by

So now you have an anti-gravity-gravity space ship, where could you go with it? It would depend on the travel time available and ship speed.

The illustration in **Fig. #33** is a guide to possible destinations within our solar system. **Fig, #33** assumes a straight line coarse at an average speed of 10,000 miles per hour (16,093 km/hr.).

As Concerns Travel Time

To meet the planet when you get to it's orbit position, depends on how direct the coarse traveled is. Starting the journey at the right time helps accomplish this rendezvous. The reason being you do not want to find yourself chasing the planet in it's orbit if you are late getting there, you may not be able to catch the planet. perhaps finding instead it is on the opposite side of the solar system from where you are. Even if you could anchor the ship in place, it could be a long wait until the orbiting planet comes around and meets you.

To get some idea of the time it takes each planet to make one orbit around the sun (termed it's sidereal period), Earth goes around it's orbit once every 365.26 days, Mars goes around every 686.98 days, Jupiter once every 11.68 years, Saturn once every 29.46 years, Uranus once every 84.0l years, Neptune once every 164.8 years, Pluto makes one orbit every 247.7 years. Those hot inner planets, Mercury and Venus take 87.9 days and 224.701 days respectively.

The planetary speeds in orbit range from 48 km/sec, for Mercury, to 30 km/sec. for Earth, to 5 km/sec. for Pluto on average. As you can see the farther from the sun the planet is the slower it's orbital velocity. A planet's orbital velocity rises and falls a bit as it moves around in it's orbit from above an average to below average due to the eccentricity of it's orbit. Aiming a space ship is like shooting a slow bullet from a moving earth to hit a moving target a long way off.

An intrepid wanderer like yourself should welcome some auspicious conditions. Here are a few you may encounter visiting your extra-terrestrial other worldly destinations.

The Sun 6,000 degrees C. (10,832 degrees F.)
Escape velocity 617.5 km/sec. (384 mi/sec.) This velocity also lets one escape the solar system. The Sun's surface gravity is 27.89 times that of earth.

Mercury –400 degrees C. (752 degrees F.) on the sun facing side and minus 180 degrees C. (minus 292 degrees F) on the space facing side, escape velocity 4.25 km/sec. (2.6 mi/sec.), surface gravity 0.38 of earth's.

Venus – 480 degrees C. (896 degrees F. more than hot enough to melt lead), escape velocity 10.4 km/sec. (6.5 mi/sec.), surface gravity 0.91 of earth's.

Earth -50 C. degrees to minus 62.2 C. (122 degrees F. to minus 80 degrees F.).surface gravity 1.0, escape velocity 11.2 km/sec.
(7 mi/sec.) acceleration of gravity 32.174 feet per second per second.

Earth's Moon –surface temperature 120 C. degrees C. to minus 153 degrees C. (273 degrees F day. to minus 153 degrees F. night) escape velocity
2.4 km/sec. (1.5 mi/ sec.), surface gravity 1/6th that on earth (0.17 of earth's).

Mars - temperature 22 degrees C. to minus 23 degrees C, (71.6 degrees F. to minus 94 degrees F.), escape velocity 5.1 km/sec. (3.2 mi/sec.). surface gravity 0.38 of earth's,

The Asteroid Belt – There are 40,000 plus asteroids, 19 have diameters one mile or greater. The asteroid Ceres has a diameter of 429 miles (690 km.). Exercise caution a few have very elliptical orbits. Escape velocity from the asteroid Ceres is approximately 0.47 km/sec. (0.29 mi/sec) that would be 1,531 ft/sec. surface gravity 0.026 of earth's.

Jupiter - surface temp minus 150 degrees C. (minus 238 F.) escape velocity 60.3 km/sec. (37.5 mi/sec.), surface gravity 2.64 times that on earth.

Saturn - surface temperature minus 180 degrees C. (minus 292 F.), escape velocity 60.3 km/sec. (20.1 mi/sec.), surface gravity 1.14 times that on earth.

Uranus - surface temp, minus 210 C. (minus 346 F.) escape velocity 22.6 km/sec. (15 mi/ sec.), surface gravity 0.91 times that on earth.

Neptune - surface temp minus 220 degrees C. (minus 364 F.). escape velocity 24 km/sec. (14.9 miles per second). surface gravity 1.1 times that of earth.

Pluto - distance from sun 4,566,000,000 miles. escape velocity 1.18 km/sec. (o.73 mi./sec.), surface gravity 0.07 of earth's, Dimly lit and very cold, minus 230 degrees C. (Minus 382 degrees F.).

Escape velocity from earth works out to: 25,000 miles per hour. For earth's moon: 5,370 miles per hour. and for escape from the sun: 1,381,346 mi/hr. (2,223,000 km/hr.). If you have a ship that can make that speed, survive the temperature and stay together, you have one of the marvels of the solar system. Needless to say a ship with an anti-gravity field for propulsion and an internal gravity field for the crew is essential for escaping the sun.

Without protection from the sun's gravitational attraction, a 200 lb. (earth weight) crew member would weigh 5,578 lbs. near the surface of the sun, not counting the asbestos underwear.

Visualizing The Pull of Gravity

A gravity-well is a way to visualize the rate of increase of the gravity force as the planet, sun or moon emitting the gravitational attraction is approached or is reduced as one gets farther from the surface.

The gravity force pulling on a space ship or satellite is greatest at the planet surface and decreases the farther one gets from the surface.
The illustration **Fig. #35 page 187** shows the weakening of the gravity force on moving away from the surface of the planet. Each time one moves one planet radius farther from the planet's gravity center the gravity force is reduced as shown.

Gravitational Force and the Inverse Relationship

Any two bodies (masses) attract each other with a force that is directly proportional to the product of their masses and inversely proportional to the square of the distance between their centers of gravity. mathematically this becomes:

Intensity at a distance = <u>**intensity at source**</u>
from source   **distance from source**
                               squared

The spheres of **Fig. #34** page 186 illustrate the inverse relationship of the surface area of a sphere and the force of gravity on it's surface

The spheres increase in size one radius ('r' of the smallest sphere) at a time.

For each one square meter of surface area on the gravity generating body, the next larger spheres have the following surface areas:

      4sq.m., **9**sq.m., **16**sq.m., **25**sq.m., **36**sq.m.,

The matching change in the gravity force on the surface of these larger sphere spheres is

$$\text{1/4}^{\text{th}}, \text{1/9}^{\text{th}}, \text{1/16}^{\text{th}}, \text{1/25}^{\text{th}} \text{ and } \text{1/36}^{\text{th}}$$

respectively per square meter of surface area compared to that on one square meter of the central sphere's surface.

In this way the gravity force out beyond the planet's surface is reduced, as the distance beyond the surface increases.

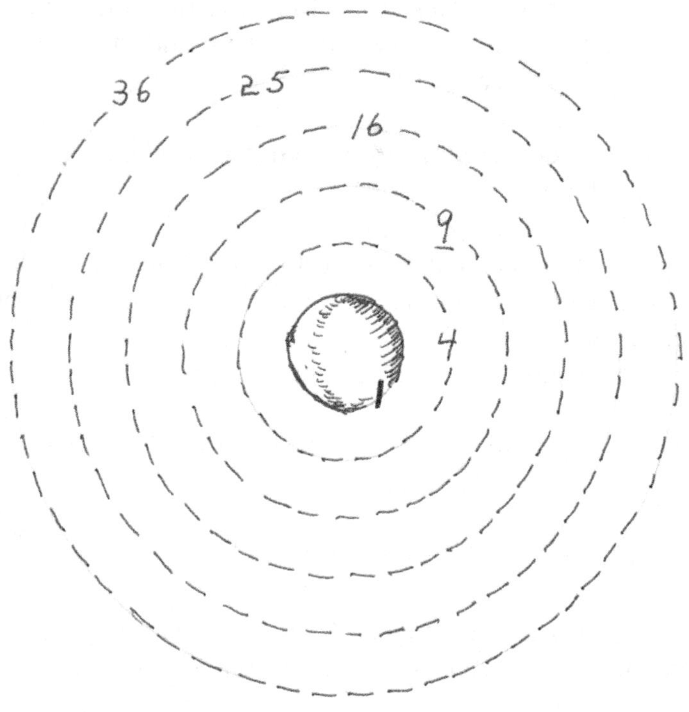

Fig. #34

The inner mass has an surface area of one, the next larger sphere an area 4 x 1, the next larger sphere an area 9 x 1, and so on. At 6 radii out the area of the sphere is 36 times that of the central body.

The Gravity Well's general shape is always the same, only the magnitude of the gravitational force changes as the mass of the central body is greater or lesser. **When the central body has a radius of "one"** The force of attraction increases as one nears the attracting mass.

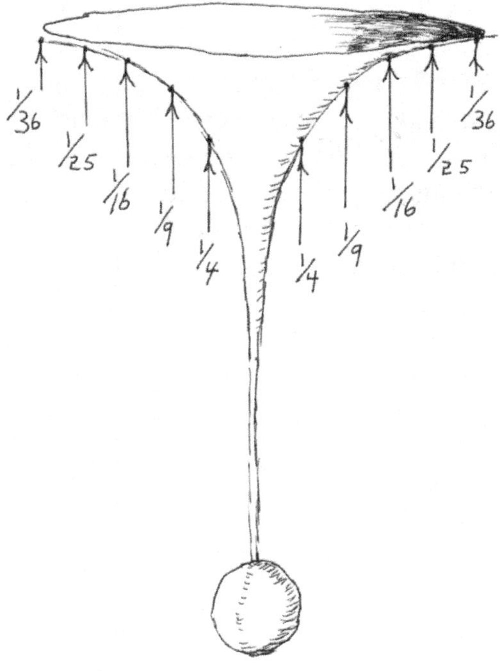

Fig, #35

The attracting mass has an radius of one. At a distance 2r the attraction force is 1/4th as strong it is at 1r. At a distance of 3r the attractive force is only 1/9th as strong, as it is on the surface, of 1r. At 6r out the attraction is only 1/36th as strong as it is on the surface of a sphere of 1r.

If you wish to learn more about the subject of space travel and how it affects orbits and satellites etc. complete with illustrations consider consulting the book:

The "Space Handbook and Analyst's Guide"
Volume II, prepared by Major Michael J. Muolo, Air University Air Command and Staff College Publication AU-18, Air University Press, Maxwell Air Force Base, Alabama 36112-6428.
ISB-N-X942, USGOV Bk-Mil, In 1999 it cost $36.00.
This book is available to the general public from any government book store. Some of the larger U.S. cities have such stores, you might consult the phone book or the internet for their locations. The U.S. printing office may also be able to supply this publication.

The "Phase" of a Wave

Phase is a comparison of a complete cycle with some feature of a cycle. All wave cycles have "Phase" relationships between the complete cycle and the three characteristics of that cycle.

If one characteristic of a cycle reaches it's peak amplitude before another, it is said to "Lead". If the characteristic reaches it's peak amplitude after that of another characteristic, it is said to "Lag" the other characteristic. See Fig. #37 page 191.

In the graphs of harmonic wave motion (see Fig. #22, page 134) note displacement reaches it's peak amplitude before velocity reaches it's peak amplitude. Therefore velocity "lags" displacement in the cycle.

Acceleration on these same harmonic motion graphs is always shown on the opposite side of the horizontal axis from displacement, directly opposite. This means that <u>when displacement is increasing, acceleration is hindering further increases</u> in displacement. This hindering ability takes the form of "strain forces" in substance, similar to elastic forces in a compressed steel spring. The strain forces are stored energy, elastic energy, potential energy, energy to rebound.

When displacement's maximum amplitude is reached, all the kinetic energy of motion of substance will have been converted into potential energy of strain force in substance, motion then momentarily stops.

<u>When displacement is decreasing, acceleration is aiding further decrease,</u> now substance is rebounding back toward it's un-displaced state, accumulated strain forces now aide substance gain velocity as it returns to zero displacement (to reach the horizontal axis on the graph). Substance on reaching

zero displacement will have converted all the potential energy of stored strain force back into kinetic energy of motion of substance (back into velocity).

The amplitudes of displacement and acceleration can be seen as complementary opposites. Acceleration and displacement as concerns phase are 180 degrees "out-of-phase". It will be 180 degrees for them to exchange positions on the graph.

Indicating Wave Phase.

Three ways to indicate a position along a wavelength or to indicate the phase of a wave characteristic compared to a reference wave. Degrees, Radians, and as Fractions.

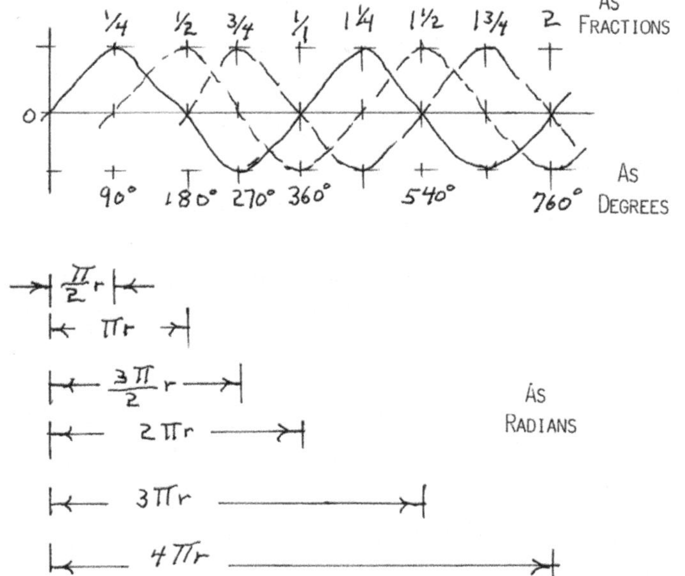

Fig. 36

360 degrees equal one full wavelength.
2 pi r   equals one full wavelength.

190

Is The "Phase" Lagging or Leading?

On comparing the phase of a wave characteristic to the reference wave, a phase shift to the right <u>of the reference wave's maximum amplitude</u> in the diagram, means the characteristic maximum begins later. Such a wave shift is said to "lag" the reference wave.

Whereas if a wave characteristic begins earlier than the reference wave maximum amplitude, it is said to "lead", and appears to the left of the reference wave. See **Fig. #37** below.

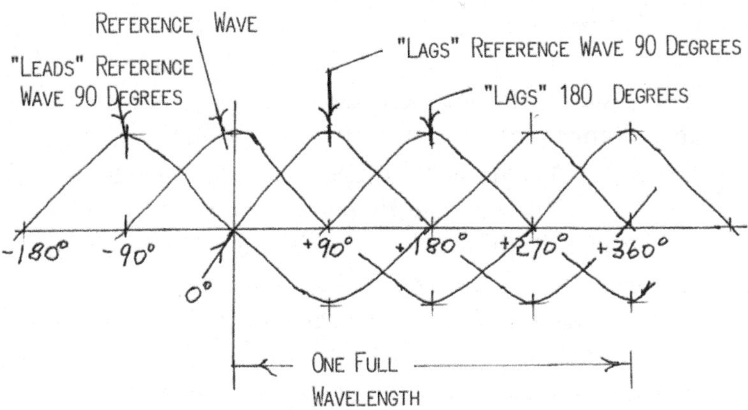

Fig. #37

Cycles In The Cosmos

The Universe, Chaos-Cosmos Cycle

In the theory in this book the universe cycles between a state of **Cosmos,** orderly patterns, predictability, balance, and harmony supreme, and a state of **Chaos,** chance, un-balance, unpredictable, non-uniform, and un-harmony reign. Back and forth the universe, rocks, between the two, first one dominating then the other. The universe itself is eternal, it's balance of orderliness and chaos ever changing each moment perpetually out-of-balance, yet over time their cycle dynamically in-balance.

Astronomer **Ernst J. Opik** (Estonia, 1893-1985) a Bruce medal holder, in his book "The Oscillating Universe" ** speculated on the whole life-cycle of the universe. He saw the existence of the universe before there were suns and planets, as a universe condensed into a super dense "**Uniform Substance**", what a physicist might term a "**Nuclear Fluid**". This super dense unified homogeneous substance would be like the atomic particles, the electron, proton, and neutron with the spaces taken out from between them. He termed this great lump, this mass, "**The Primeval Atom**", "**The Cosmic Egg**". The egg from which the universe we have come to know, was born. The Cosmic Egg was the product of a previous contraction of an existent universe, the elastic forces with the Primeval Atom were tremendous, and once the velocity of contraction slowed and stopped the Primeval Atom virtually exploded, expanding outward, this expansion, it is speculated is continuing to this day. Today it is millions of years since the Primeval Atom began expanding.

* "The Oscillating Universe" by Ernst J. Opik,
 c 1960, a Mentor Book published by The New American Library of World Literature Inc. U.S.A. library of Congress catalog card #60-9595.

Opik estimates that the expansion of the visible universe is proceeding at about 1/5th light speed. Far into the future, millions of years from now the expansion will slow and stop, then the contraction part of the cycle will begin again, eventually condensing into a new Primeval Atom. Opik termed this eternal expansion and contraction of the universe **"The Cosmic Pendulum"** or **"The Oscillating Universe"**.

Opik stated that the complete cycle of expansion and contraction would take 30,000 million years, 15,000 million to expand and 15,000 million years to contract. He speculated that the expansion has presently been going on for some 4,500 million years since the cosmic egg exploded.

Whether there was a cosmic egg that exploded or not I leave to the experts. As far as the theory in this book is concerned the universe cycle is one of chaos to cosmos for one half and cosmos to chaos for the other half an endlessly repeated cycle, an eternal cycle. There may well be a cosmic egg of sorts at some period in this cycle.

The Vast Immensity, The Milky Way Galaxy

It's dimensions are often stated in "Light Years" to get some sense of how far light can travel in one year in the vacuum of interstellar space, it is **5,878,000,000,000 miles** (9,459,724,000,000 Kilometers, or in scientific notation 9.46 x 10 to the twelfth power kilometers). Astronomers are talking thousands of these light years when describing the Milky Way galaxy.

A 360 degree color photograph has now been made of the Milky Way Galaxy, it is a beautiful piece of work showing the brilliant nucleus of the galaxy, dust clouds, and so many stars astronomers can only make and educated guess as to their number (over 200 billion). The outer diameter of the spiral arms is from 200,000 to 300,000 light years, The heavily populated galactic disc has a diameter of 120,000 light years and the densely populated galactic bulge or Nucleus has a diameter of 12,000 light years. The whole immense Milky Way is not just sitting still, it is whizzing through space at an estimated 550 kilometers per second (342 mi/sec.) in relation to nearby galaxies, it is headed in the general direction of the Hydra Constellation and the Virgo Galaxies. You need not be afraid the Milky Way will crash into another galaxy soon as the nearest one is some 30,000 light years away.

The Milky Way galaxy is rotating and completes one revolution every 225 million years, our Sun held in one of it's arms is carried along with that rotation in a huge elliptical orbit around the galaxy center. Our sun is at a distance of approximately 27,000 light years from the galactic center moving at a speed of from 210 km/sec. to 240 km/sec. (130 mi/sec. to 149 mi/sec.).
Electromagnetic emissions of stars in the Milky Way galaxy range from the extremely short gamma and X-rays, through all of the visible spectrum, the infra-red (heat waves) and on down into down into longer and longer waves such as radio waves.

Other names for the Milky Way

In Latvia, Lithuania, Finland and Estonia it is "The Bird Path" or "The Way of the Birds", in China it is "The Silver River", in Wales in the Welch language it is "Caer Wydion" which is "The Fort of Gwydion", in Iceland and Norway it is "The Winter Way", and in Japan it is "River of Heaven".

Dimensions Of The Milky Way Galaxy

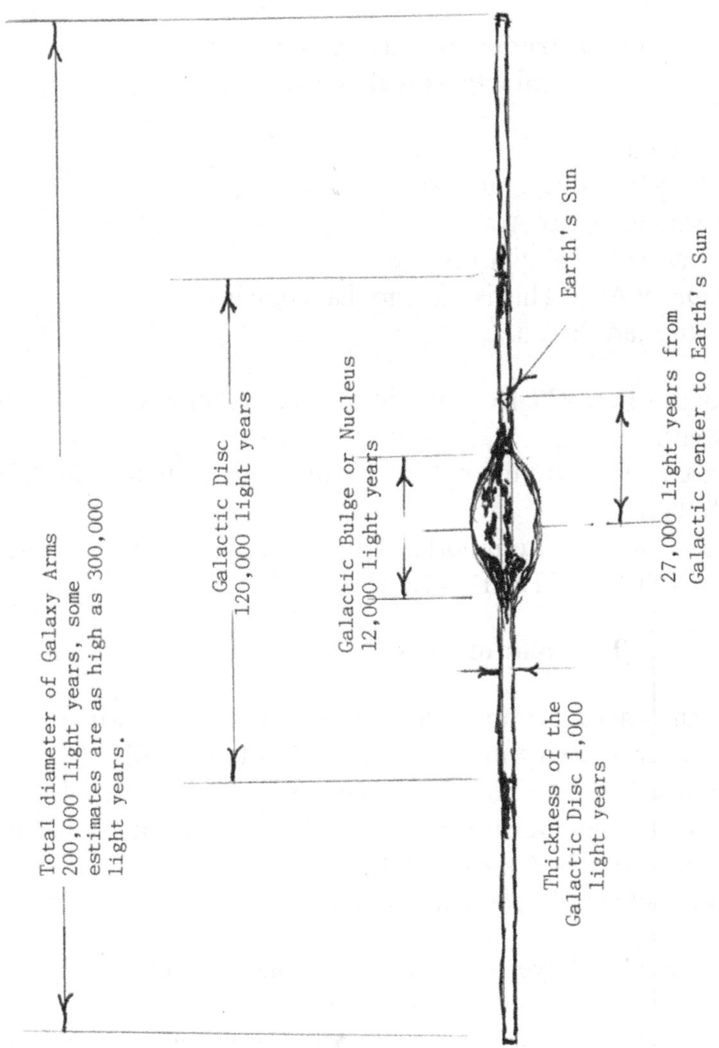

Fig. #38

The Spiral Arms reach out from the galactic disc to the perimeter of the Milky Way galaxy. There are four main arms and two smaller arms

**Names astronomers have given names to
the galactic spiral arms.**

1) Perseus Arm
2) Norma and Cygnus Arm
3) Scutum, Crux Arm
4) Carina and Sagitarius Arm
5) Orion Arm, This is the arm that contains
      the earth's sun.

Note in **Fig. #39** Deneb is in the constellation Cygnus (The Swan).
Crux as seen from south of Earth's equator is termed "The Southern Cross".
Polaris as seen from north of Earth's equator is termed "The North Star" or "The Pole Star".

The color of stars

To the naked eye most stars appear a points of white light, however a few of the brightest have a tint of color to them. **Antares** appears reddish (a sign of a relatively cool star), **Rigel** appears bluish-white (sign of a relatively hot star), and some stars like **Capella** and our own sun have a yellowish tint. Photographs show many stars to have distinct colors.

To the naked eye here are some stars that appear brightest (the smaller the number the brighter).
Sirius -1.58D, **Canopus** -0.86, **Alpha Centauri** 0.06D, **Vega** 0.14, **Capella** 0.21, **Arcturus** 0.24, **Rigel** 0.34D, **Procyon** 0.48D, **Archernar** 0.60, **Beta Centauri** 0.86, **Altair** 0.89, **Betelgeuse** 0.92. Four of these are double stars indicated with a 'D'. One is a variable brightness star (**Betelgeuse**).

The Milky Way Galaxy

seen from Earth which is in a spiral arm

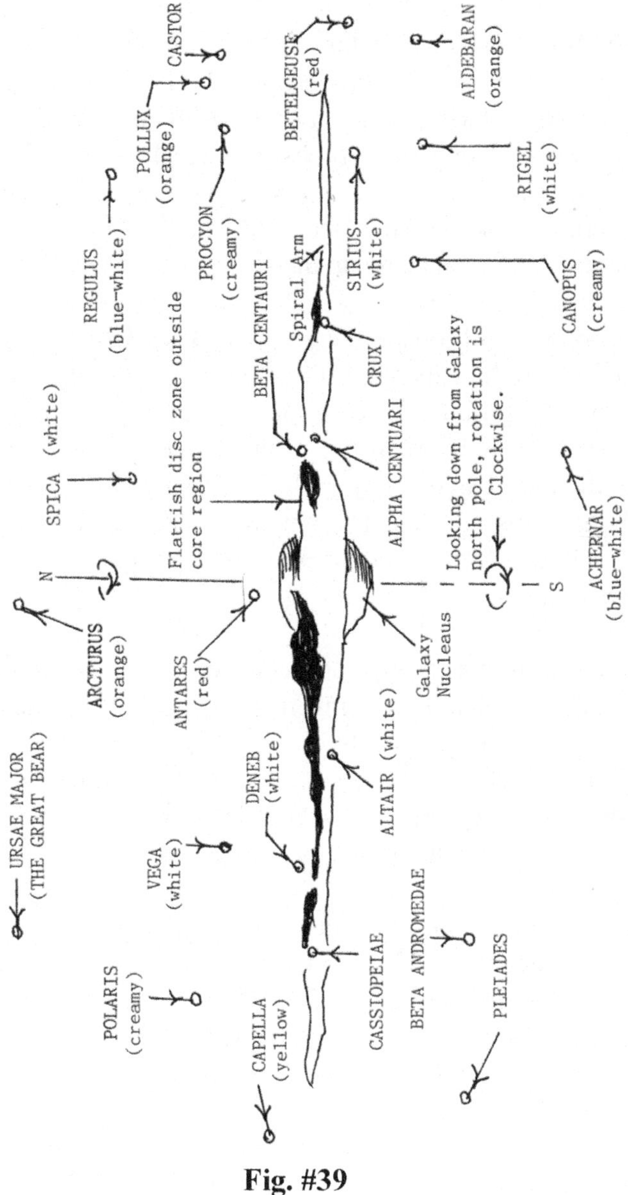

Fig. #39

The Rotation Of the Milky Way Galaxy

Our Milky Way galaxy takes about 225 million years to make one revolution. below are presented the more recent geologic eras of earth. During these eras the Milky Way galaxy rotated about two and one third times. On Earth during these eras great shallow seas existed and dried-up, mountain chains rose and fell, and some time about ten million years ago large carnivores and man came on the scene.

Roughly 570 million years ago the **Paleozoic era** began. The Paleozoic era had six ages, The first age of the Paleozoic era, was the Cambrian age, then came the Ordovician age (480 million years ago), then the Silurian age (435 million years ago), then the Devonian age (400 million years ago), then the Carboniferous age began about 340 million years ago, and the final age of the Paleozoic era, the Permian age began 260 million years ago.

As far as living things are concerned it was during the **Paleozoic era** that marine invertebrates, fishes, needle bearing trees (conifers), gingko trees, tree ferns, and club mosses flourished, early reptiles appeared, insects abounded including the sharp sighted Dragonflys, and sharks swam the warm seas. This is the era when coal deposits were laid-down.

After the Paleozoic era's last age the Permian age ended a new era began **the Mesozoic era**. The Mesozoic era's first age was the Triassic age (225 million years ago) the age of dinosaurs, then came the Jurassic age (180 million years ago when birds, and the first small mammals appeared, mammals are warm blooded creatures as opposed to reptiles which are termed cold blooded, since reptiles do not make their own heat, but depend on the warmth of their environment as do insects.

Then came <u>the Cretaceous age</u> (began 130 million years ago), It was during the Cretaceous age that grasses and cereal grains appeared, the extinction of dinosaurs occurred, and the U.S. Rocky Mountains arose.

About 70 million years ago the **<u>Cenozoic era</u>** began with **<u>the Tertiary Age</u>**. During the Tertiary Age placental mammals, flowering plants, whales, apes, and <u>perhaps</u> early man appeared.

Then came the age we are in now, **the Quaternary Age** (began about 10 million years ago), and man came on the scene. Yes man learned to make and use tools, draw pictures on cave walls, develop a writing system, but unfortunately man seems to have great difficulty mastering himself, to the determent of his environment and every living thing in it.

The glaciers have advanced and receded four times during the Cenozoic era. In 2010 the glaciers are receding again and quite rapidly due to global warming, they may disappear altogether in a few years.

All the above things have occurred during two and roughly one third revolutions of the Milky Way galaxy, the galaxy we call home.

that have taken place <u>during one 225 million year revolution</u> of the Milky Way galaxy, Only part of the carboniferous age took place during this revolution.

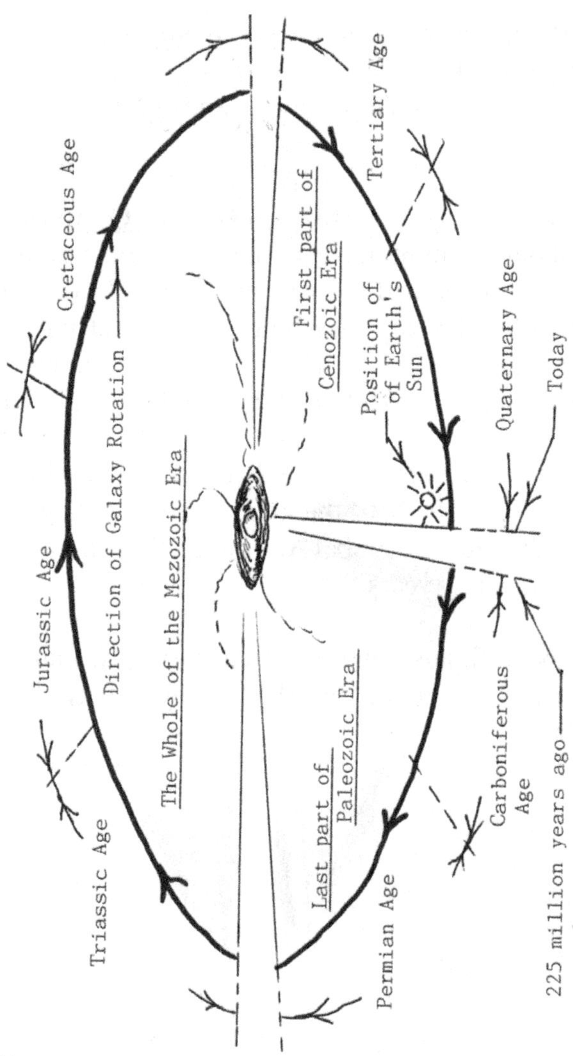

Fig. #40

Note On Astronomical Distances

If you viewed the original "Star Trek ". series, you may have heard Spock ask, what is our present location?

The answer was something like that which follows. Captain we are 2 parsecs from our original position.

How far a distance is a parsec? Astronomers also at times use "parsecs" as a unit of measure. A parsec is 19.2 trillion miles, or 3.26 light years, or 206,265 times the radius of the Earth's orbit, or if you really want astronomer talk, a parsec is a distance having a heliocentric parallax of one second.

If the star ship Enterprise could travel two parsecs <u>in half an hour</u>, they would have done so at a speed of:
2 x 2 x 19.2 = **76.8 trillion miles per hour**.

It is not the speed, astounding as it is that would bother me, it is the stopping when we get there that could be a bit of a bother.

The closest star to earth in the Milky Way galaxy is **Alpha Centauri** at 4.3 light years away. In the constellation Taurus is the star group **The Pleiades "The Seven Sisters"** (the seven stars of this group visible to the naked eye), they are estimated to be roughly 100 parsecs (300 light years) from earth. Stars in another galaxy are much more distant, the closest one is probably in the Andromeda galaxy (Messier catalog #M31), at 2 1/2 million light years away. It is reasonable to use a faster than light ship to visit such places. A ship that could <u>average</u> light speed would shorten the time to visit objects in our solar system as well. At light speed it would be one and one third seconds to the moon, 8 minutes to the sun, the sun to Jupiter in an hour, the sun to Pluto in 5 1/2 hours.

The Elliptical Orbits of Planets and Moons

The Earth Orbiting The Sun

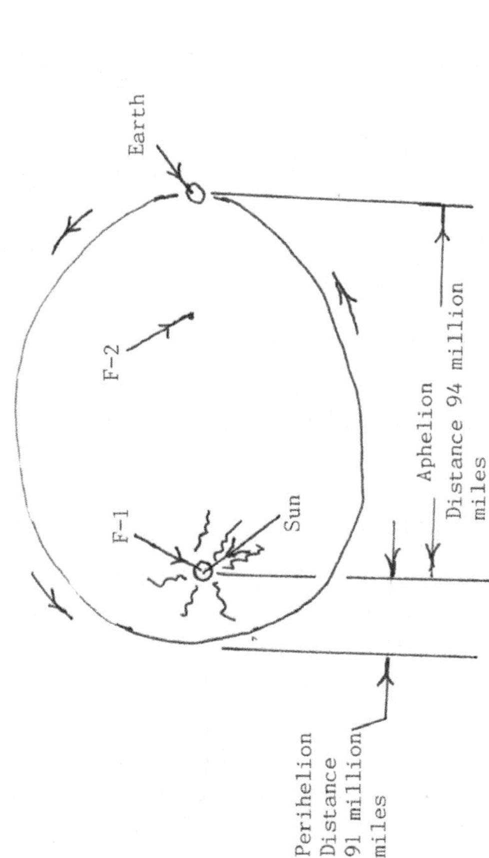

Fig. #41

The Moon Orbiting The Earth

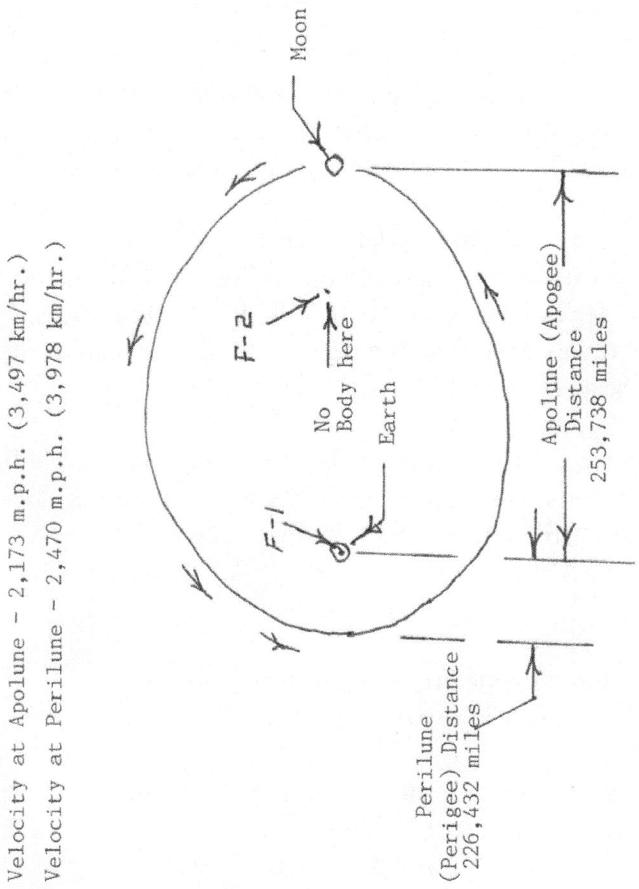

Fig. #42

All planets and Moons have elliptical Orbits in our solar system. At one end of the ellipse the orbiting body pivots **F-1 quickly** on a short radius, then it travels to the other end of the ellipse and pivots **F-2 slowly**, on a long radius.

The Simple Harmonic Motion of Earth Moving
In It's Orbit around the Sun

Earth travels in an elliptical orbit around the Sun taking 365.24, days to make it around it's orbit once. This single cycle around it's orbit is termed a "Sidereal Year".

The square of the sidereal period of any planet is proportional to the cube of "mean" distance from the Sun. The "mean" distance for Earth is 92,957,200 miles. Note the Earth travels fastest when closest to the Sun and slowest when farthest from the Sun.

The mass of the Earth is roughly 6600 million, million, million tons. Mean surface gravitational acceleration of rotating Earth is 32.174 feet per second.

The Simple Harmonic Motion of the
Moon in orbit around the Earth

The Moon travels in an elliptical orbit around the Earth, it takes 27.32166 days (27 days 43 minutes 11.5 seconds) to complete one orbit (termed the sidereal period).

For the Moon to go through it's phases it takes 29.53 days (29 days 12 hours 2.8 seconds), (termed the synodic period). The Moon's "mean" distance to Earth is 238,840 miles.

Mass of the Moon is 0.01227 that of Earth. Note the Moon travels fastest when closest to Earth and Slowest when farthest from Earth.

How would an Elliptical Orbit Change
a Harmonic Wave Graph?

Firstly I am not a mathematician, so my readers who are mathematicians will need to prove for themselves the two graphs that follow represent accurately the harmonic wave motion of a planet in an elliptic orbit.

I begin with the fact that the motion of a planet in orbit is a regularly occurring motion. The planet oscillates, that is resonates in it's orbit, between having a fast velocity at the perihelion end of the ellipse and a slow velocity at the aphelion end of the ellipse.

In short the motion in orbit is harmonic wave motion that must have the three wave characteristics, displacement, acceleration and velocity, and as such can be graphed

For a circular orbit all is fine and good, see **Fig. #43 graph page 207.** The orbit is symmetrical with the center if gravity (F-1 and F-2 would fall at the same point).

For an elliptical orbit, see **Fig. #44 graph** page 208. The center of gravity is closer to one end of the ellipse than the other. To adjust the graph for this I moved the zero neutral horizontal axis downward on the graph as needed.

The consequence of this moving of the horizontal axis is that it shows the planet takes more elapsed time to negotiate the orbit one side of the center of gravity (the aphelion side) than on the other (the perihelion side). Is this really the case? I believe it is.

Aside from their straight forward amplitudes on the graph, the three characteristics represent more, as follows:

Displacement – Amount of displacement in orbit from center of gravitational attraction.

Acceleration – The strain in the substance of space. The **Potential energy** of the moving body, stored energy, acceleration hinders or encourages a change in the rate of motion. The potentiality of change.

Velocity – The stress in space, the force that "molds" substance, that moves substance, it represents **kinetic energy** of the moving body Kinetic energy is the actuality of change.

So look over the two graphs and see what you think. I have never seen planetary motion described in this way in any text, theory or mathematical paper. Visualizing planetary motion as harmonic wave motion is one more way to understand what is occurring.

Circular Orbit

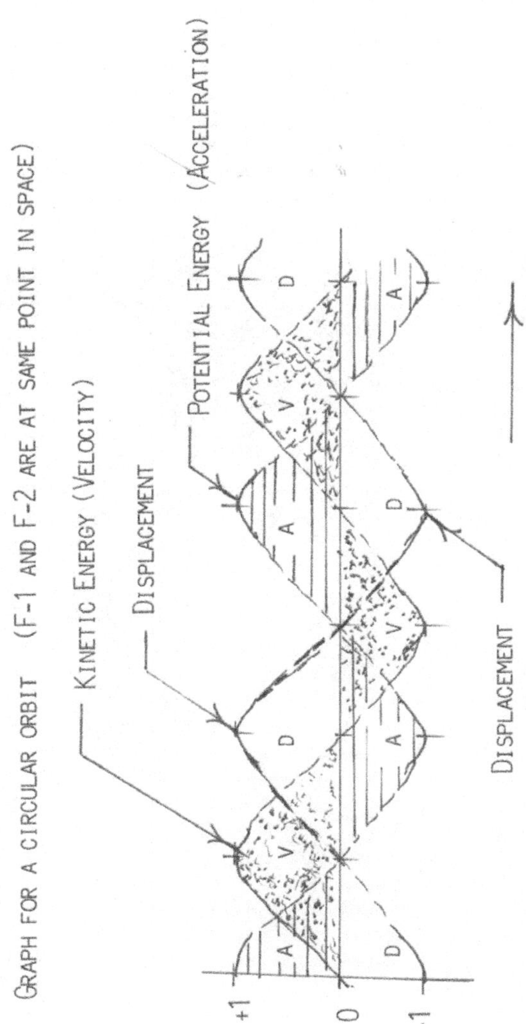

Fig. #43

207

Elliptical Orbit

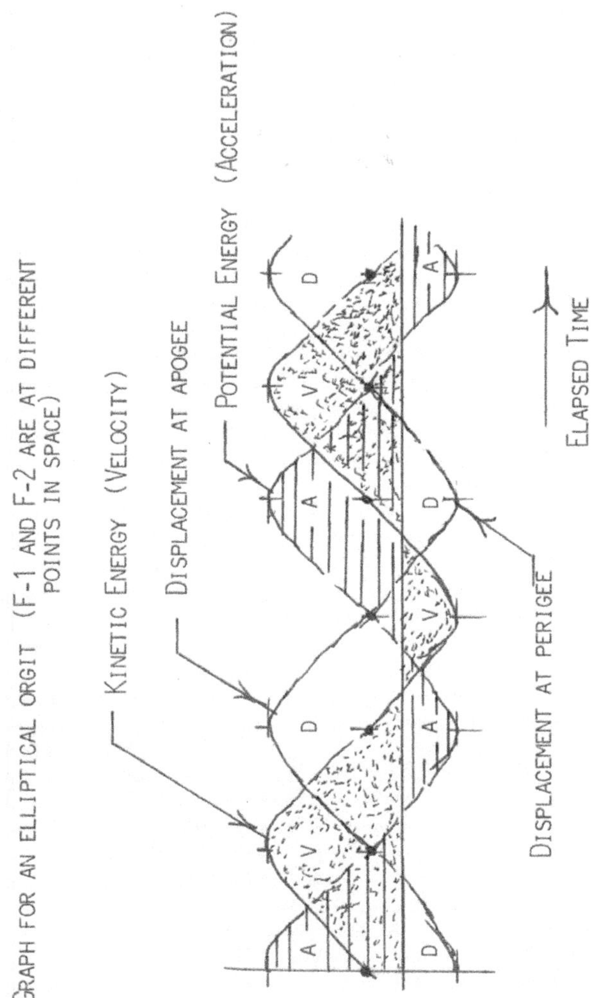

Fig. #44

Brief Concluding Comment To "I Am Universe"

This book evolved from a basic premise:

> **"What 3,500 years of man's**
> **experiences with the universe**
> **has made him feel about it**
> **that man, has expressed in**
> **writing."**

I added my own thoughts to the speculations and conclusions of the sages in the form of a theory that is compatible with many of their ideas.

I hope you can use what I have written like a tool, to explore this reality we find ourselves in that we term "Universe".

Sincerely,

E. R. Margis

Appendix A

Exponential Notation and the SI

The Conference Generale des Poids et Mesures (CGPM) a body responsible for all international matters concerning the metric system, named their approved system of units

"Systeme International d'Unites"

Usually abbreviated to

the SI system

The system uses decimal arithmetic. Units of different sizes are formed by multiplying or dividing by a base of ten.

An example of this with measures of length is:

10 millimeters = 1 centimeter
10 centimeters = 1 decimeter
100 centimeters = 1 meter
1000 meters = 1 kilometer

In such a system changes from one of these powers of tens units to another is made easily by shifting the decimal place over.

thus 10.0 millimeters becomes
1.0 centimeters.

and 1,000.0 millimeters becomes
1.0 meters.

A Table of the **SI** prefixes for these multiples follows.

Table of SI prefixes symbols and multiplication factors

	Multiplier	Prefix	Symbol
The very large	10^{24}	yotta-	Y
	10^{21}	zetta-	Z
	10^{18}	exa-	E
	10^{15}	peta-	P
	10^{12}	tera-	T
	10^{9}	giga-	G
	10^{6}	mega-	M
	10^{3}	kilo-	k
Larger	10^{2}	hecto-	h
	10^{1}	deca-	da
Smaller	10^{-1}	deci-	d
	10^{-2}	centi-	c
	10^{-3}	milli-	m
	10^{-6}	micro-	μ
	10^{-9}	nano-	n
	10^{-12}	pico-	p
	10^{-15}	femto-	f
	10^{-18}	atto-	a
The extreemly small	10^{-21}	zepto	z
	10^{-24}	yocto-	y

These multiplier prefixes can be applied to any SI unit, some of which are:

for length the metre - m
for mass the kilogram - kg
for time the second - s
for electric current the ampere - A
For temperature degrees Kelvin - K
for amount of substance the mole - mol
for luminosity the candela - cd

For a positive exponent the decimal place is moved the indicated number of decimal places to the right. 1.0000 times ten to the third power would = 1000.0

For a negative exponent the decimal place would be moved the indicated number of decimal places to the left. 1.0000 times ten to the minus third power would = 0.001

Appendix B
Poetic wisdom from the past

In a piece of poetry written about 300 B.C. by a now unknown author(s) that were concerned with the value of human life, we find the extremes of human behavior, all driven by the "Forces" of "Need" and "Satisfaction".

"All things have their season, and in their times all things pass under heaven. A time to be born and a time to die. A time to plant, and a time to pluck up that which is planted. A time to kill, and a time to heal, a time to destroy, and a time to build. A time to weep, and a time to laugh. A time to morn, and a time to dance. A time to scatter stones, and a time to gather. A time to embrace, and a time to be far from embraces. A time to get, and a time to lose. A time to keep, and a time to cast away. A time to rend, and a time to sew. A time to keep silence, and a time to speak. A time to love, and a time of hatred. A time of war, and a time of peace."

Ecclesiastes ch.3.,vs.1-8

The author goes on to say what he believes the best of all change driven by need-and satisfaction is in a person's life. "And I have known that there was no better thing than to rejoice, and do well in life."

Ecclesiastes ch.3,vs.12

The Earthly realm, the realm of the "Now", where change is continuous and life so ephemeral, so transient. The poet comments.

> "Man born of woman, living for a short time, is filled with many miseries. Who cometh forth like a flower, and is destroyed, and fleeth as a shadow, and never continueth in the same state."

> Job ch. 14, vs. 1-2.

and again as:

> "For we are sojourners before thee, and strangers, as were our fathers. Our days upon earth are as a shadow, and there is no stay."

> The Book of Chronicles 1,
> ch. 29, vs. 15. *

* "Chronicles has also been titled:
Paralipomenon (#1 and #2). The Hebrews call this book "The words of the days" (Dibre Haijamim). Chronicles is thought to have been written about 400 B.C.

Appendix C

Introductory Explanation of
a "Matrix Diagram"

What is a matrix Diagram? It is a way to illustrate the relationship of three basic components and how their four main characteristics interact.

This is accomplished using a short family tree arrangement, attached to a matrix of nine squares. This, nine square matrix is like the one used in the game "tic-tac-toe".
The family tree begins with the term the whole diagram originates from. The family tree then progresses to downward to where this singularity term divides into two distinct complementary opposites, and from their, moves on down into the matrix diagram itself. The nine square diagram has <u>four main characteristics,</u> each is located <u>in one of the corner squares of the diagram</u>. The squares between the corner squares are terms that result from the inter-action of the corner square characteristics.

Some of the diagrams introduce the concept of what have termed **"Planes of existence"** or **"World Systems"**, an Aryan-Hindu concept which in the theory in this book consists of, **the four parts of a cycle of change**, plus one plane that is the unified whole of the four parts of the cycle of change. This is located in the center square.
The location of the five planes of existence, are as follows: Each side of the matrix diagram (has three squares) is one of the four planes, which are **Atma, Buddhi, Kama,** and **Sthula**.

<u>The fifth plane of existence, is at the top of the family tree, it is termed **Paran-atma or Brahman,**</u> it is special in that it's essence is the entirety of the diagram. Brahman is eternal

existence, and as such is un-born, unlimited in extent or diversity. At the destruction of order in the universe Brahman alone is awake (active). Brahman is Free Will, Chance, disorder, symmetry, unlimited Substance and Energy.

"Ishvara" is Brahman taking forms, Brahma with limitations. Ishvara is all that is born Ishvara is the inner controller of born things. Ishvara regulates the form, and life span of created things. Ishvara is probability, asymmetry, and order.

The Value of a Matrix Diagram

1) The family tree shows the emergence of duality from singularity, and the connectedness that links the dualities. This connectedness is "The Messenger" that lets one duality affect the other, and contra-wise.

2) A matrix diagram shows what I believe were once termed the "Pillars" or "Supports" of the Universe, these as interpreted in this theory are the four parts to a cycle of change, The "worlds" of the Aryan-Hindus.

3) The nine square matrix diagram shows the main characteristics of two of the basic components and how they interact.

Constructing a Matrix Diagram

Three partial matrix diagrams will now be shown as an aide to constructing a new one, they will be followed by five complete matrix diagrams, which are:

1) The Physical Universe. 2) Metaphysical Reality. 3) The Mythological Universe. 4) The Mind (Manas). 5) The Psychological "Self".

Diagram A: Shows "The Totality", the Universe as either chaos or cosmos at the top of the family tree, below it the emergence of duality is shown, Below the duality are shown the four main characteristics of the duality.

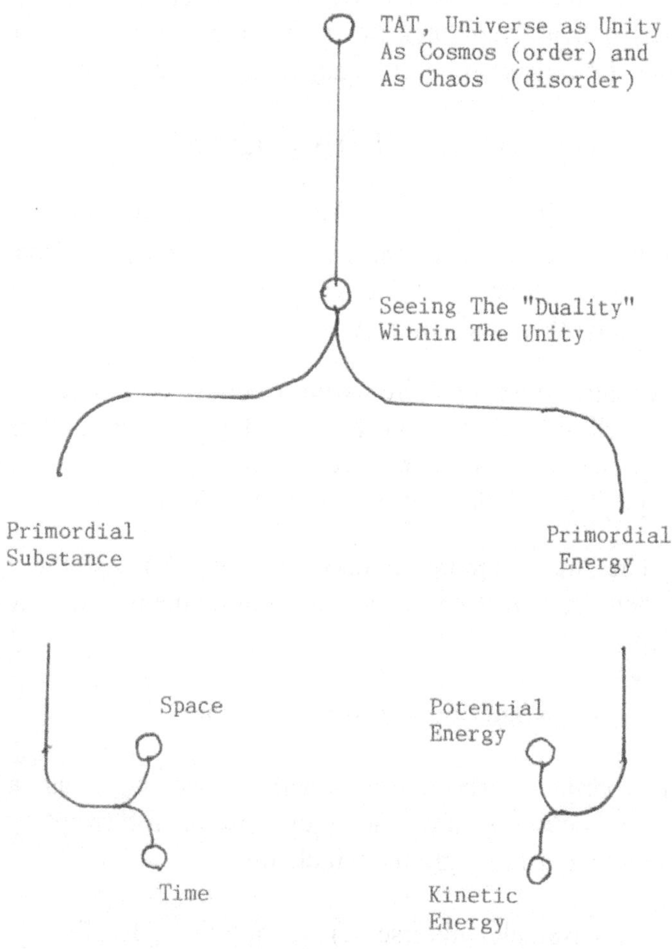

Diagram A

Diagram B: Here is shown the link between the Complementary opposite dualities. Actually this connecting link emerged at the same time as the duality pair, but is shown here in a separate diagram for clarity.

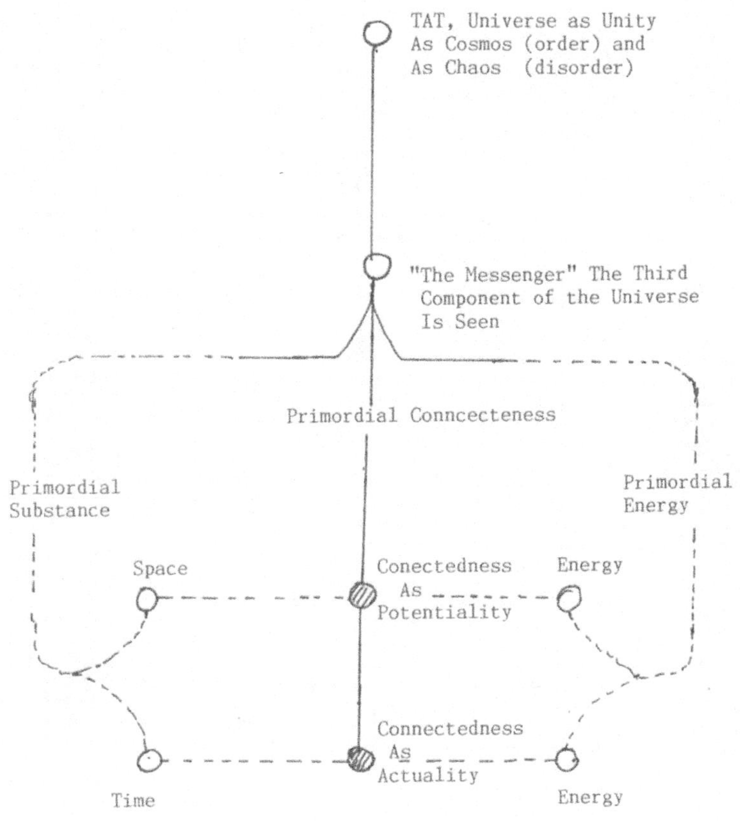

Diagram B

Diagram C: In this diagram one can see which squares of the matrix diagram have terms that are **potentialities** (top horizontal row) and which terms are **Actualities** (bottom horizontal row). The center horizontal row contains terms that are resultants of the interaction of terms in adjacent squares.

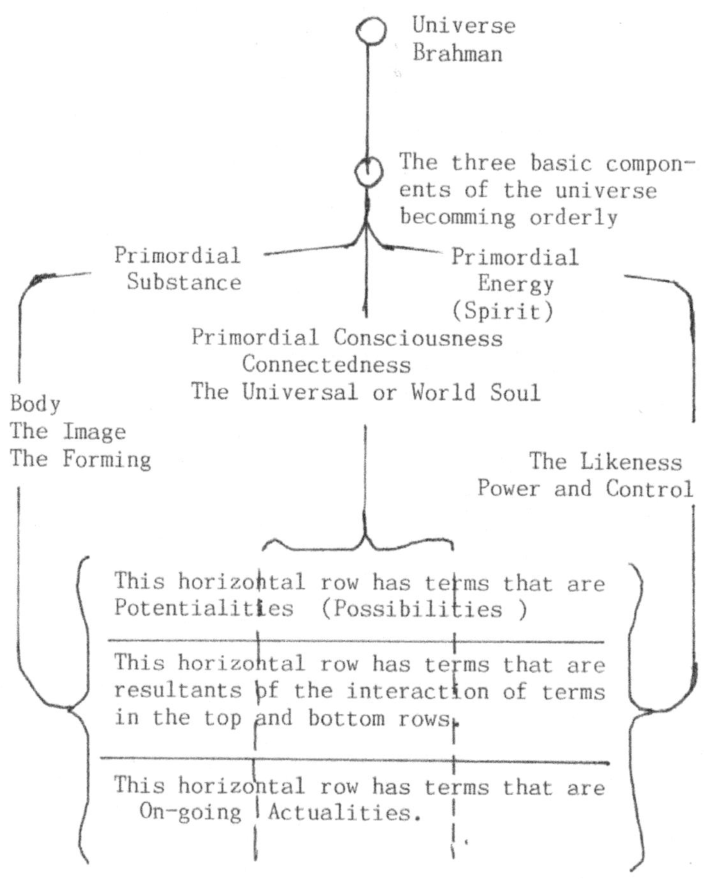

Diagram C

The section to follow offers five complete "Matrix Diagrams"

Diagram #1a The Physical Universe

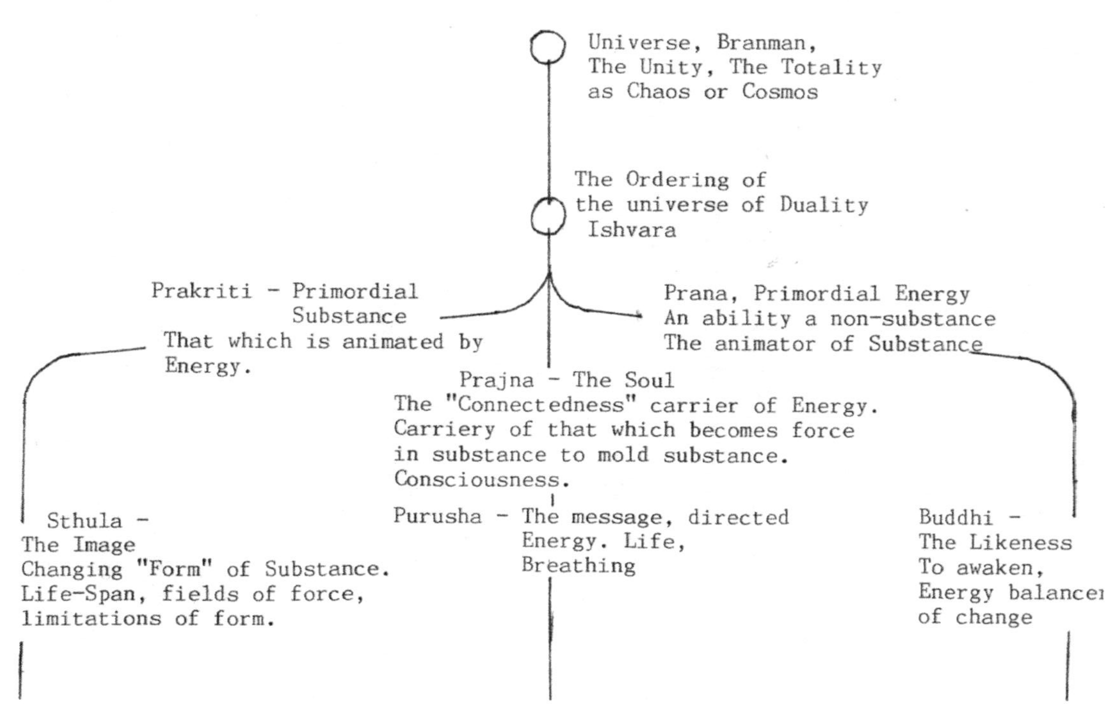

220

Diagram #1b The Physical Universe

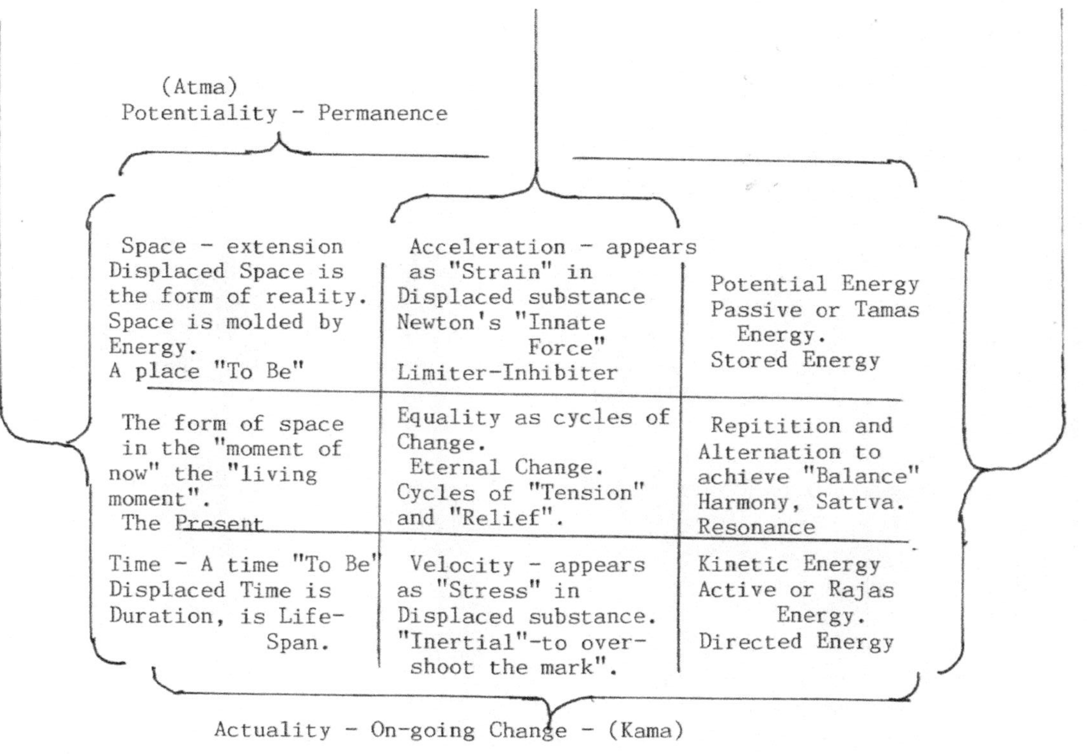

221

Diagram #2a Metaphysical Reality

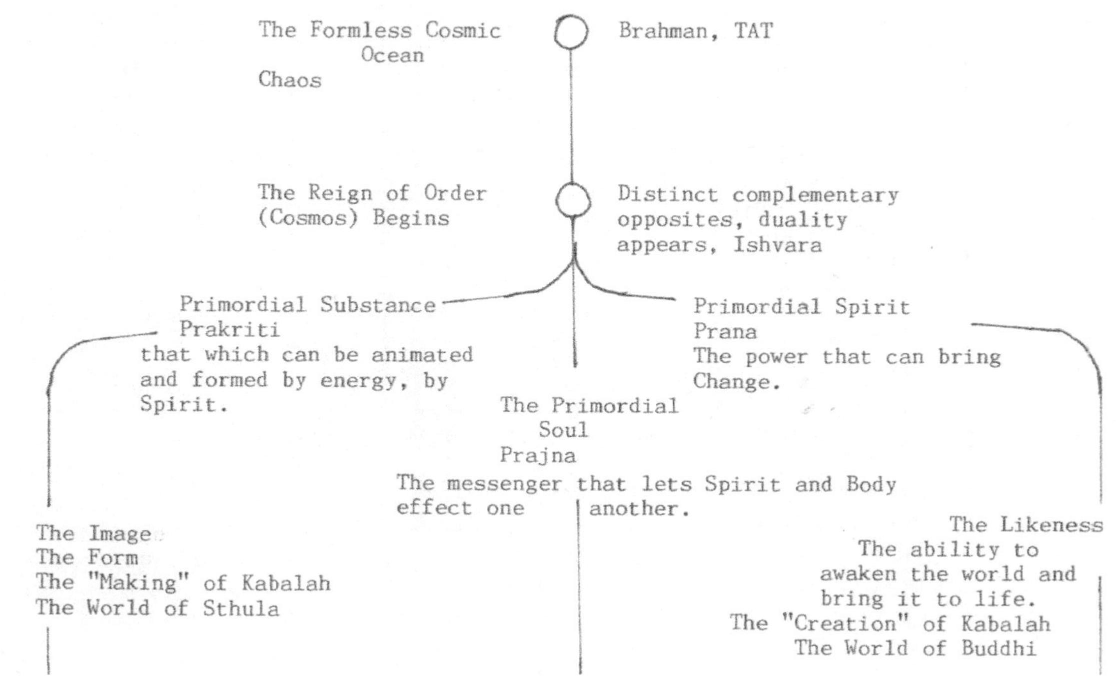

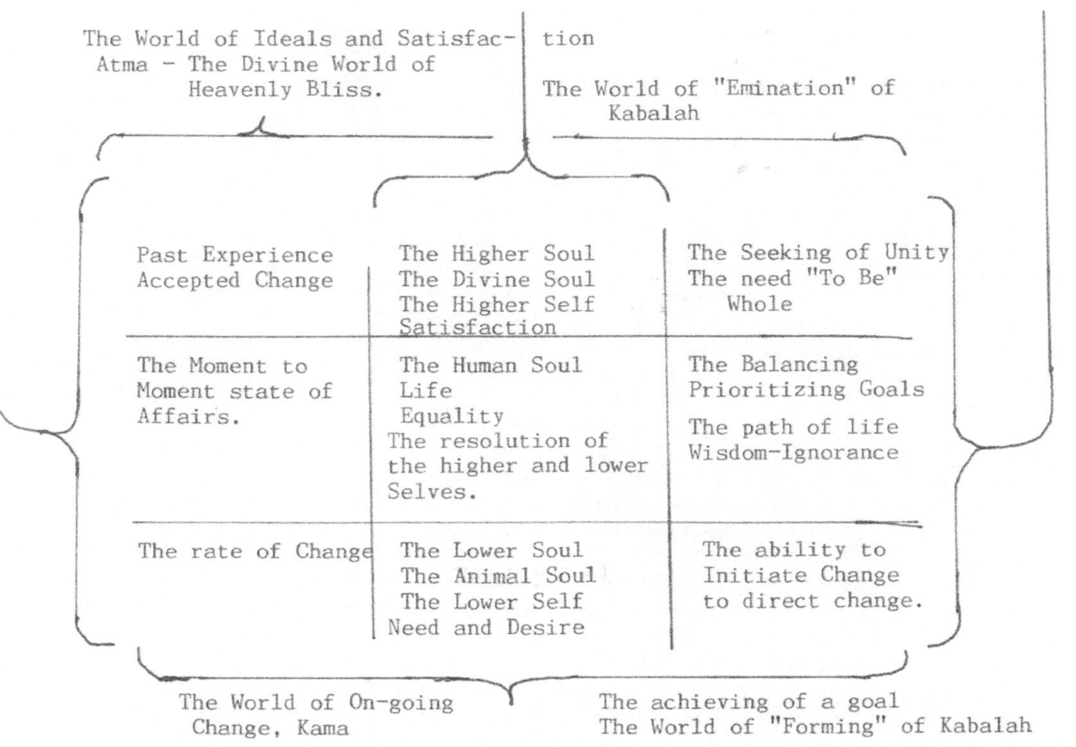

The World of Ideals and Satisfac-|tion
Atma — The Divine World of
Heavenly Bliss.

The World of "Emination" of
Kabalah

Past Experience
Accepted Change

The Higher Soul
The Divine Soul
The Higher Self
Satisfaction

The Seeking of Unity
The need "To Be"
Whole

The Moment to
Moment state of
Affairs.

The Human Soul
Life
Equality
The resolution of
the higher and lower
Selves.

The Balancing
Prioritizing Goals
The path of life
Wisdom-Ignorance

The rate of Change

The Lower Soul
The Animal Soul
The Lower Self
Need and Desire

The ability to
Initiate Change
to direct change.

The World of On-going
Change, Kama

The achieving of a goal
The World of "Forming" of Kabalah

Diagram #3a Mythological World

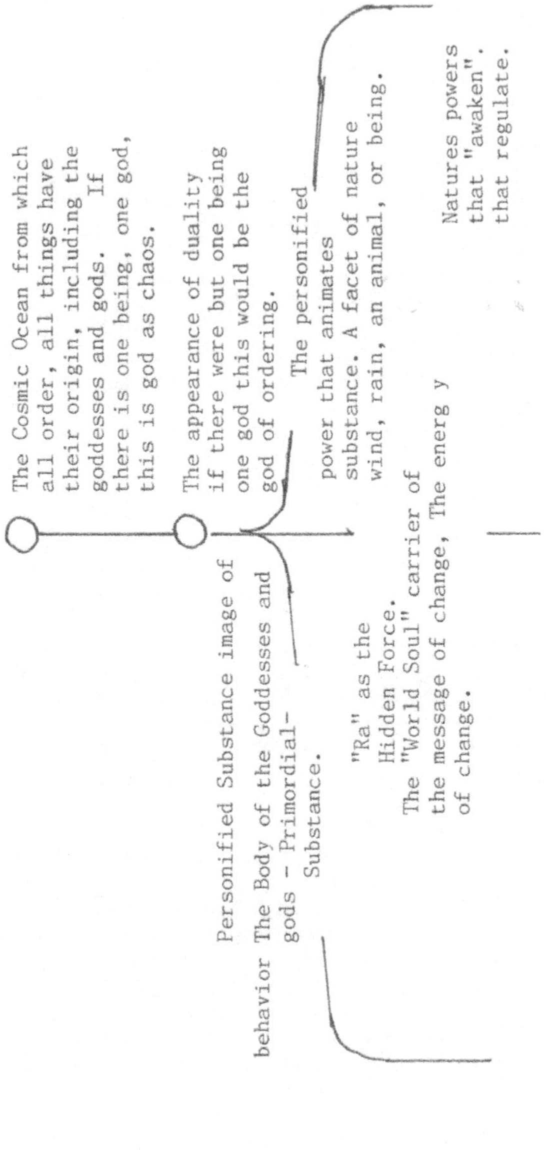
The Cosmic Ocean from which all order, all things have their origin, including the goddesses and gods. If there is one being, one god, this is god as chaos.
The appearance of duality if there were but one being one god this would be the god of ordering.
The personified power that animates substance. A facet of nature wind, rain, an animal, or being.
Natures powers that "awaken". that regulate.
Personified Substance image of behavior The Body of the Goddesses and gods - Primordial- Substance.
"Ra" as the Hidden Force.
The "World Soul" carrier of the message of change, The energ y of change.

Diagram #3b The Mythological World

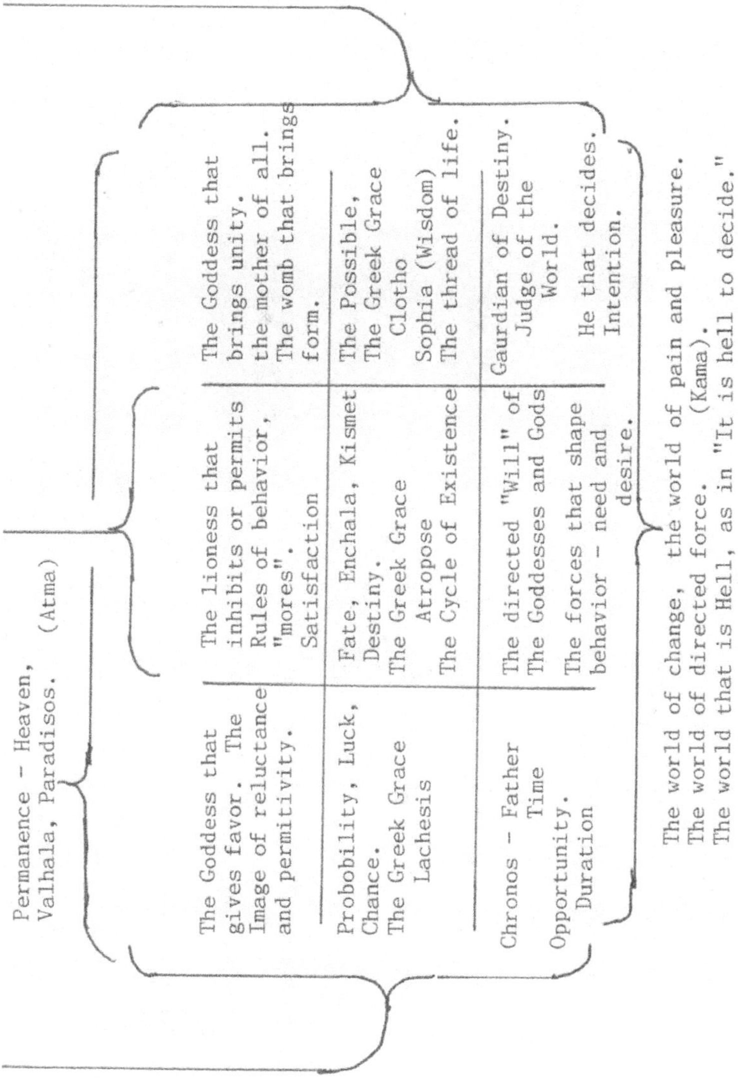

Permanence – Heaven,
Valhala, Paradisos.   (Atma)

The Goddess that gives favor. The Image of reluctance and permitivity.	The lioness that inhibits or permits Rules of behavior, "mores". Satisfaction	The Goddess that brings unity. the mother of all. The womb that brings form.
Probobility, Luck, Chance. The Greek Grace Lachesis	Fate, Enchala, Kismet Destiny. The Greek Grace Atropose The Cycle of Existence	The Possible, The Greek Grace Clotho Sophia (Wisdom) The thread of life.
Chronos – Father Time Opportunity. Duration	The directed "Will" of The Goddesses and Gods The forces that shape behavior – need and desire.	Gaurdian of Destiny. Judge of the World. He that decides. Intention.

The world of change, the world of pain and pleasure.
The world of directed force.   (Kama).
The world that is Hell, as in "It is hell to decide."

Diagram #4a Mind (Manas)

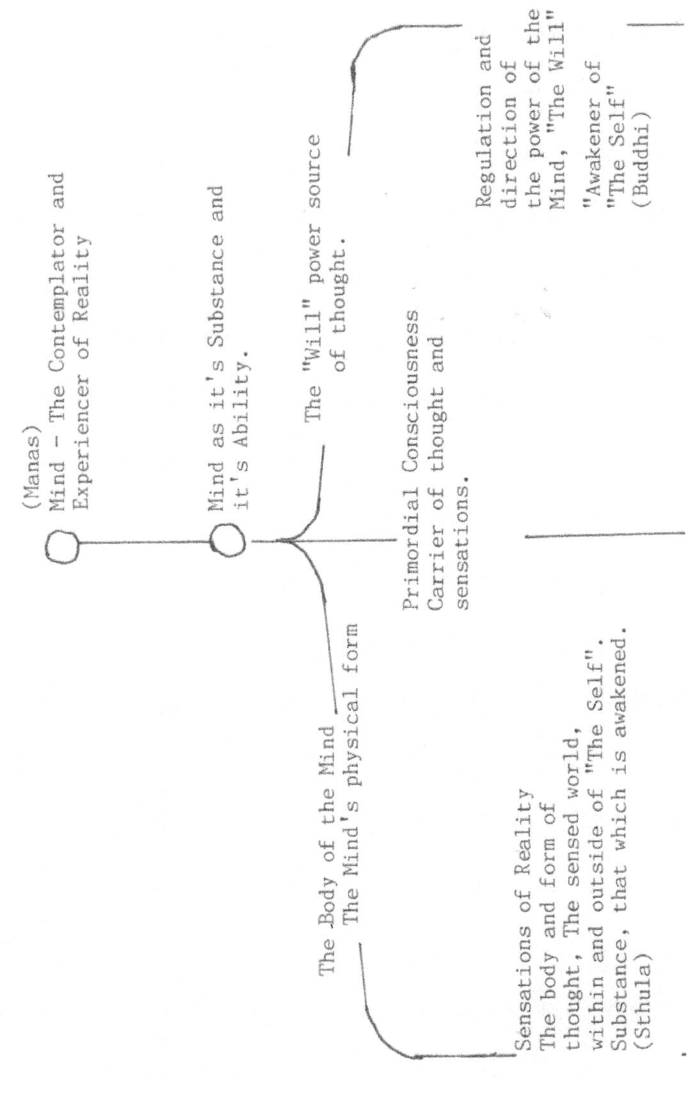

226

Diagram #4b Mind (Manas)

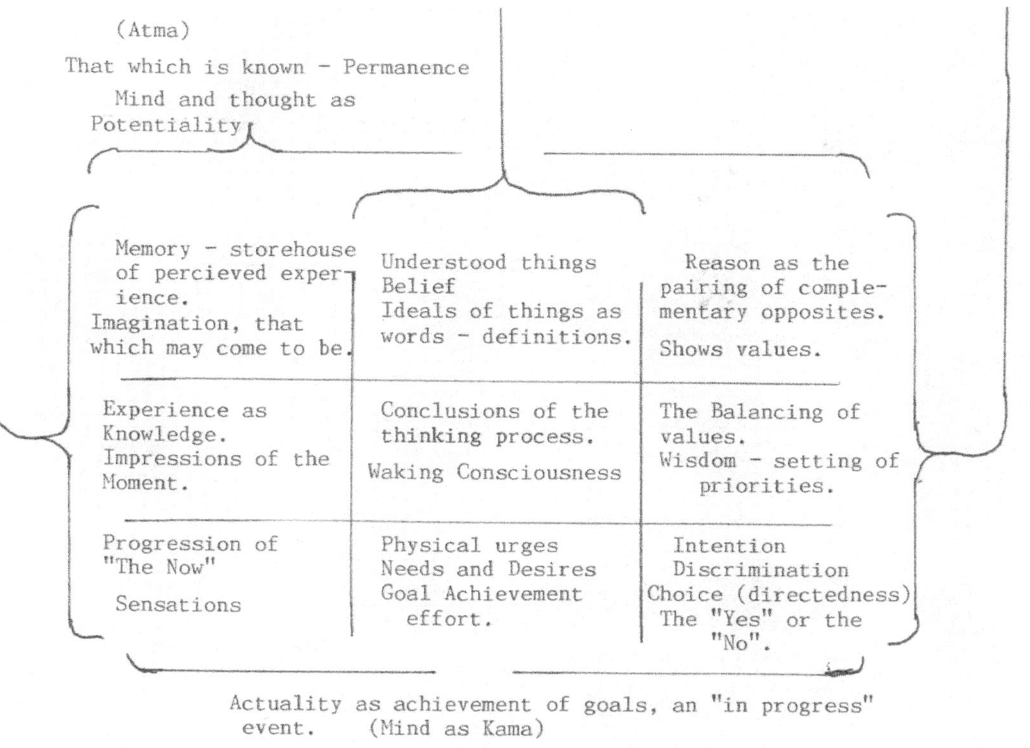

The Psychological "Self"

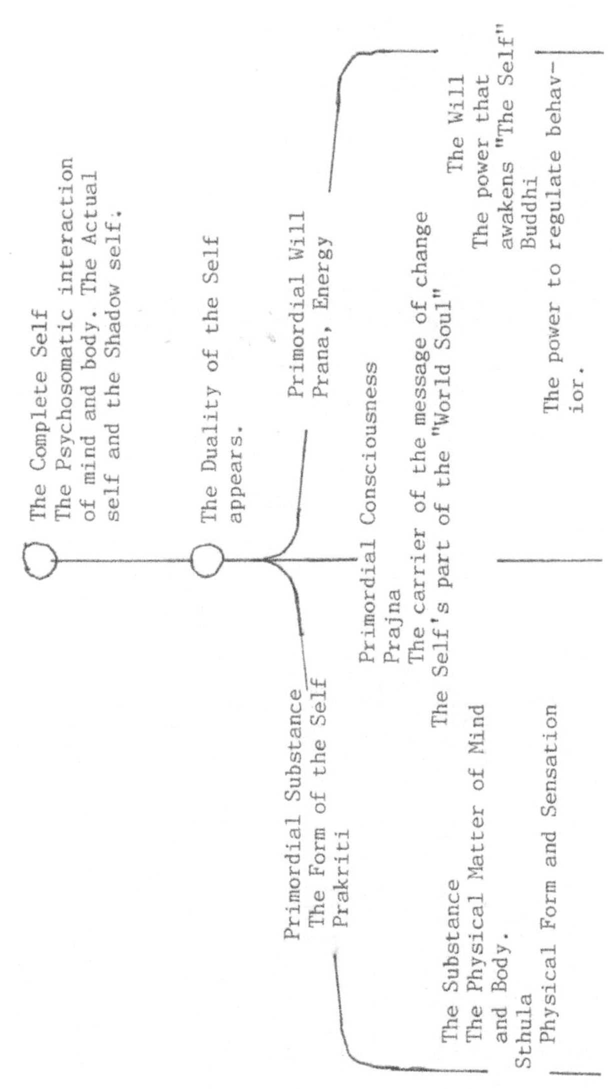

Diagram #5a

The Psychological "Self"

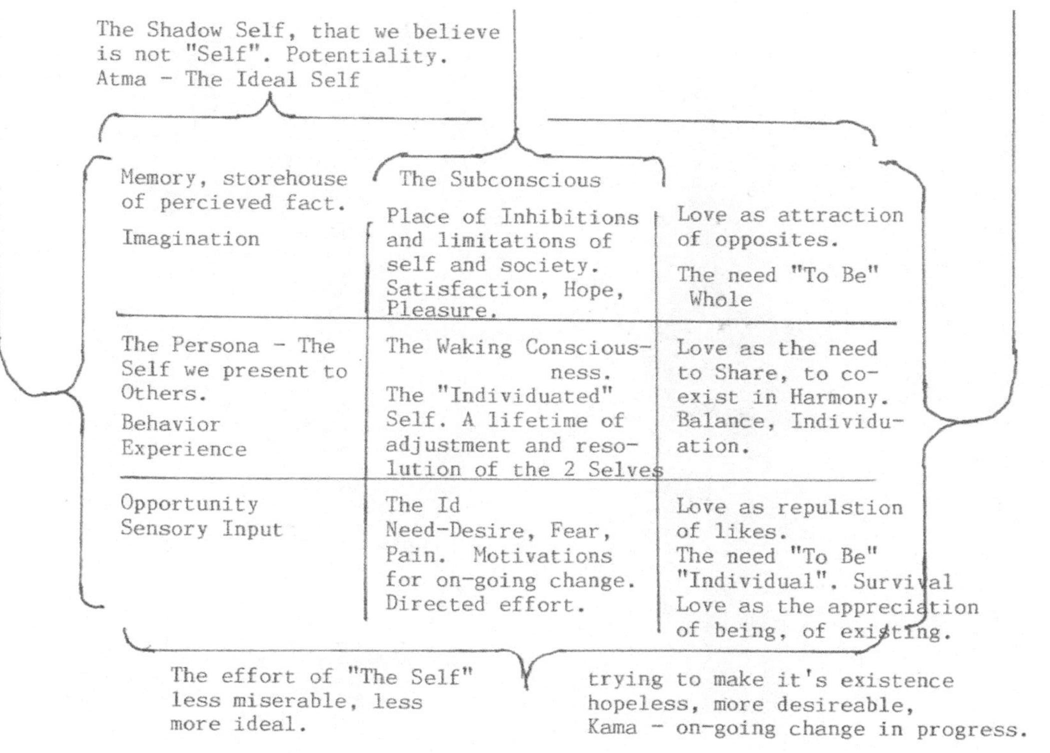

Diagram #5b

The Two Selves of Man
Me and My Shadow

Fig. #45

The "Two Selves" of Man

Psychologists, study behavior, and from their studies have come to believe there are two selves in a person. The two selves are:

"The Self we believe we are"

and

"The Self we believe we are not"

The Self we believe we are not they term,

"The Shadow Self"
the self that is "Other"

Psychologists now realize a person of either sex has both behaviors within, their sexual gender and their role in life may favor one, the masculine with its aggressiveness and physical strength for a male, or the feminine with it's gentleness and motherliness of the female. However there are times when the male needs to be gentle, and the female, needs to show strength; balancing the two behaviors is essential, a man that is all macho or a woman that is all fluff can have a problem reacting to, or interacting with others.

Behavior of the "Actual Self" of the moment must be balanced with the behavior of the "Shadow Self", the "Potential Self".

A person's aim in their conduct is to balance the two selves and as a result of so doing, "get along" with others around them in a reasonable understanding manor. Psychologists term this process of balancing the two selves "Individuation". Individuation is a life long process, a life long effort. I see this balancing as essential, each person needs a measure of control in their life from moment to moment, and some freedom as far as is practical to choose their destiny, meaningful freedom to choose, a sense of control in what is happening in ones life.

The Two Selves of the ancient Egyptians

The Egyptians at the time of the pharaohs did not believe the mummified bodies of the dead would rise and live again, they believed if a person
lived a life that was "worthy", their soul would be permitted to rise to the realm of the gods and exist there eternally. If they were not "worthy" when they died, there heart, which they
believed was seat of their emotions would be destroyed, and their soul would be placed in a new body, and they would have to live an earthly life of struggle, suffering and need, all over again.

Egyptians at the time of the pharaohs also believed a person had a twin self of the opposite sex that lived on a twin earth. This may have begun as someone musing that since Egypt was a kingdom of two lands, perhaps a person had a twin living in the other land.

Modern psychologists carried the idea a step farther and put both twin selves, "The Self", and "The Shadow Self", in one body.

The metaphysical way of understanding the reality of man is, the whole man had two parts, **"The Higher Self"** and **"The Lower Self"**.

The "Higher Self" is an eternal self, a self that has all it needs, and all it desires satisfied, it is a perfect self.

The **"Higher Self"**, exists in a world of "Ideals", a world in which satisfaction and bliss are located. The higher self is a **"Potential Self"**, it is the reality of the potential.

The **"Lower Self"**, metaphysically exists as **"Actuality"**, it is an on-going incomplete self. The lower self is the way things are with the person "right now". and is always in "need", "desirous" of something, existing in a world where itself and others can make errors, can be hungry, be harmed, be unloved, and be unheard.

The upper self could not be harmed, made no error. The Upper Self only notes that which is error. The higher self exists in the realm of the goddesses and gods, and like them the higher self is eternal. Perhaps the higher self is at one with Brahman, at one with that which is, "complete knowing", complete consciousness, and complete power. The **"Higher Self"** exists as **"Potentiality"** to be the complementary opposite of the actualized Lower Self. I would point out the higher and lower selves are part of a complete whole, <u>**one without the other cannot exist alone**</u>, one without the other is useless.

The Matrix Diagram and the two selves of man

In the psychological matrix diagram of "The Self" the top horizontal row, of three squares, is the potential self, the "Higher Self", the self that reacts, and the bottom horizontal row of three squares is the self of "Actuality", the "Lower Self", on-going directed change. The lower self's primary concern is survival of the physical body, and the concerns of the physical body. The higher self's concerns are those of "perception", "belief" (understanding), and "Reason". The higher self interprets information from the senses, interprets the world of experience.

The world of the higher self is freed from the cares of the body, lives in a state of bliss, a state of complete satisfaction.

Depending on which view of the two selves of man one cares personally to subscribe to, each has a truth to it.

Man has the self of his physical body with it's sexuality, and a second self that is concerned with sorting out the world of it's experience, then reacting to it as it sees fit.

The Lower Self, The Physical Self, will eventually perish, but the higher self in a way can live on in others who were influenced by the actions, words and deeds the lower self carried-out whilst the person lived.

Question:

Do personified goddesses and gods with gender (femaleness or maleness), have a "Shadow Self"?

I do not ask this question idly for if my placement of Brahma the god of creation is correct (see page 266), "he" is actually a "she", a goddess. If Brahma has a shadow self it leaves open the possibility that Brahma even as a she when called upon could have an assertive side, that is could act with the characteristics of a male under certain circumstances just a woman's "Shadow Self" allows her to so do.

The speculation that Brahma is female is not out of place. The old testament book of Genesis I believe originally said creation was done by the 'Elohim.
The 'Elohim made man, The 'El part means a deity, this is followed by '**ohim**, which in Hebrew indicates the deity is both feminine and plural. In essence what the text said was, **The goddesses made man,** which makes excellent sense as no male ever gave birth to a child. Translators later altered the wording to fit a new view they wished to promote.

It has been asked "Does Love Have a Purpose?"

Over the years a lot has been said about "Love" and opinions vary on just what "love" consists of. Psychologically, that is from a behavioral point of view, I am of the opinion that there are three modes or types of "Love". **The three ways to "Love" are:**

Love as "the need to be whole" (seeking the unity of opposites),

Love as "the need to be individual" (the survival and care for the self), and

Love as "a sharing equally in existence" (which is love as a balance, as "Sattva").

Long ago now a rather well known Frenchman of his time I believe answered the question "Does Love Have a Purpose". He was **La Rochefoucauld** (1613-1680, a classical author, Duc de Francois La Rochefoucauld), his comment on love is ageless.

"Le plaisir de l'amour est d'amier"

(French)

"The Pleasure of love is in the loving"

(English)

Love is life, and if so it can also be said Life is Love.

Notes

Appendix D

Conclusions on the Nature of Waves

1) Traveling waves
  a) Do not interact with one another due to primordial substance having no mass.
  b) The displacement of primordial substance causes "strain" in the substance, an elastic force that varies in amplitude with the amplitude of displacement.

2) Spherical Standing Waves in Primordial
Substance.
  a) Are originally generated by a rare simultaneous coincidental meeting in space (a coming together) of just the needed multiple traveling wave conditions to form a spherical standing wave. After formation the formed wave is self resonant, it's oscillation self supporting.
  b) Only Spherical standing waves, have the property of mass due to their in-place pulsations. These waves are also the source of the gravitational force.
  c) These waves probably fall into dissolution by a reverse process similar to that which formed them.
  d) These spherical standing waves can interact with traveling waves in that they can reflect traveling waves, and in some instances reradiate them at a different frequency.
3) As noted in standard physics texts, mass (spherical standing waves) can interact with other masses.

In a traveling wave or spherical standing wave there are no energy losses, only shifts in the balance of kinetic and potential energy, possible frequency change, and reduction in overall amplitude by the inverse square rule, when they spread out from their point of origin.

237

Notes

Glossary

word definitions as used in this book
to help avoid misunderstanding

Acceleration - Acceleration is the rate of change of velocity. Velocity is the rate of motion in a specified direction. "Speed" has no specified direction. Acceleration either limits or encourages on-going change. As the displacement of substance in this theory is increasing or decreasing. Acceleration slows the increase down, just as an automobile climbing a hill slows down on the up-side, and increases speed, on the downside. Acceleration is the reaction to change.

Active Energy (Kinetic Energy). The energy of motion, energy of action, as compared to passive energy which is a reaction to change.

Alter Ego - The second self. In some cultures the alter ego is seen as a trusted friend, in others as a guardian spirit. In psychology the alter ego, may well be "The Shadow Self".

Anti-Nodes - These are places (over time) of maximum fluctuations of displacement amplitude. In a sound wave, at a displacement anti-node air pressure rises to above atmospheric then falls to below atmospheric pressure. The horizontal axis on a graph of displacement is the median of the two extremes, it is at atmospheric pressure. A displacement anti-node in a standing wave is always located at the closed end of an air column, or the free end of a rod or bar.

239

Atma - One of the six world systems or planes of existence of Aryan-Hindu metaphysics. The six are Paran-atma, Atma, Buddhi, Kama, Sthula, and Manas (mind). In this book Atma is the world of the Goddesses and Gods. It is the world of perfection, of Ideal forms, of permanence. Atma is a world of definitions of things, and of reason.

Body - The volume of substance. A "body" has shape, form, and a character to it. A body may be living or non-living. The body of the universe can be seen as it's formed substance, it's character driven by the energy it posses.
A body can also be a non-material thing such as a whole idea, the complete form or shape of the thought comprising an idea or concept.

Belief – A state of mind in which something perceived is thought to be true. This perceived truth may be a physical fact, activity, or statement received. A belief can be modified or disregarded if sufficient reason, evidence, or circumstances prove that belief untrue.

Brahman - In this theory Brahman is the "the universe", as cosmos (order) or as chaos (disorder). Brahman, the eternal, the imperishable absolute reality. Brahman, the entirety of all that exists. It can be said "All is Brahman". Brahman is not to be confused with Brahma one of the triad of Hindu gods, (Brahma creator, Vishnu the maintainer, and Shiva the destroyer). The goddesses and gods are Brahman as "Ishvara", as limitation(s).

Buddhi - Means "To Awaken". In this theory it is the world of energy, and of energy regulation. The Aryan-Hindu world of Buddhi, includes, wisdom, reason, discrimination, and decision (the yes or no).

Consciousness – To have a consciousness of something is to sense it, to experience it, to realize it.

Consciousness is a person's awareness and perception of their existence, awareness of the physical sensations of being, awareness as Mental activity, emotions and concerns about the situation at hand. Consciousness is awareness of one's self and the world outside of "self". "**The Stream of Consciousness**" would be a running conscious state over a period of time. In the theory in this book, The universe as a whole is conscious of all it's parts and their states, and conversely each part feels the effect of the whole on itself. Consciousness is also termed **"Soul"**, or **"The Connectedness"** in this book.

Compassion – Is to realize "I and you are the same being" or as **Cicero** (Cicero, Marcus Tullius, 106-43 B.C. Roman statesman, scholar, writer) put it in De Amicitia "A friend is as it were a second Self".

The Cosmic Ocean - Source of all substance and all energy. This is the **Apsu** of the Sumerians, and the **Nu** or Nun) of the Egyptians at the time of the pharaohs, the TAT of the Aryan-Hindus, and the universe of this theory.

The Counter Earth – When the Pythagoreans spoke of the counter earth, they were, referring to earth's moon.

Cycle - A regularly recurring sequence of events. Also a recurring period of time such as the day or week, or the motion of a planet in it's orbit that is repeated over and over. A simple Harmonic wave in this theory has three cyclic characteristics, Displacement, Velocity, and Acceleration.

Description - a statement of the properties of a thing or it's relationship to other things serving to identify it.

To Desire - To want something earnestly to hope, to crave for something, as to wish to be loved. To "long" for relief from some affliction. To crave knowledge.

A Deity – is a personification of the basic powers, characteristics, or components of the natural world. A deity is termed a god or goddess. Brahman seen as limitation or things with boundaries, is termed "Ishvara". All goddesses and gods are "Ishvara". Brahman that is not termed "Ishvara" is unbounded, eternal, and without limitation.

Diffraction – is the result of a wave front when encountering a barrier with a small opening in it or a slit, this will cause a new set of semi-circular waves to be generated on the opposite side of the opening, centered on the opening. This diffraction phenomena occurs with all waves, electromagnetic, sound waves, water waves in fact it can be seen in a gap at a breakwater, outside the breakwater the waves repeatedly bump into it, on the protected side of the gap repeated semicircular waves will radiate from the gap toward the shore.

Displacement is to remove from it's proper place and to move it to another place.
In a sound wave displacement is the motion of air particles, Light wave motion was once thought to be motion of a mythical Aether substance. In this theory, displacement is the change of position of primordial substance <u>from an original aligned position</u> to an <u>un-aligned position</u>, this is in the form of a torsion motion of moveable space substance compared to space's co-existing unmoving counterpart as described in this book's text.

Duration - A period of Time. Any period of time can be said to have or be "a duration", A state of lastingness. A duration is a continuation of existence of something over a period of time, over a duration.

Ego - the inner mental self as contrasted to the physical inner self. Ego is "self esteem", self motivation, the metal self as the determiner of behavior. Ego controls thoughts and actions. At it's most basic ego is concerned with survival, self preservation.

Elasticity – The capability of a deformed body (substance) to recover it's original shape after being deformed. Springiness.

Energy - The ability to animate substance. Energy is an ability to cause and maintain change. The definition scientists like is, energy is the ability to overcome a resistance and do work. In old myths, the words power and energy were used interchangeably.

Energy of Motion, Kinetic Energy is the motion of material bodies or of primordial substance, the power to mold substance, to deform substance. In the theory in this book, kinetic energy is velocity that "molds" the substance of the universe.

Stored Energy is Potential Energy. Potential energy is the condition of substance in a displaced (strained) state. Potential energy is energy due to position, composition, or state (solid, liquid, gas, or plasma).

Equality - To be "Equal" is to be the same in some way. The "Equality" can be an equality present during an instant of time, or one achieved over a period of time.

Example, If two pumps are pumping water at variable rates at any one simultaneous instant, there outputs in Liters per minute may never be "Equal" however over period of a day, each pumps and equal number of Liters.

When the word "Equality" is placed in the center square of a matrix diagram, it means a change with this "dynamic equality" to it. It is the type equality possessed by the change occurring over a complete Chaos to cosmos cycle, or a potential energy to kinetic energy exchange cycle.

Force - a push or a pull. Force is needed to change the speed or direction of a body (mass), force is the only cause of such change. Forces occur in pairs, force and counter-force (action force and re-action force). The effect of a force on a body is to change it's speed, direction of motion, or to deform the body's substance.

A googolplex - a googol is the figure one, followed by 100 zeros. (1 x 10 to the hundredth power) **a googolplex** is the figure one, followed by a googol of zeros (Mathematically notated as: 1 x ten to the tenth power taken to the hundredth power.).

History - A listing, telling or recording of past events, or relationships. The chronological listing of persons, groups, activities, interests, etc. The listing is a "History" of such things.

A history may be actual fact based, or one of perceived fact (what one believes occurred), or be imaginary that is fictitious. In this book any historical events referred to are believed fact based, however men write history and men are fallible.

The Id also termed "**The Un-Conscious**". The Id is the body's urgings for change. The Id is that part of consciousness concerned with bodily survival, with instinctive actions such as the "flight or fight response", pulling a hand out of a fire automatically, etc. The Id of our mind "knows" the level of bodily discomfort or pleasure, "knows" the longings of the body as "needs" and desires.

Ideal Forms - In this theory the most ideal form is the "thing" itself, however in the mind a most accurate word definition is used to represent the thing, this is termed it's "ideal" and substitutes for the actual thing.

Image - In the biblical book of Genesis, Man is said to have been formed in the "**image**" and "**likeness**" of God. In this theory, **Image is the "form"**, the shape, the behavior of a thing. "**Likeness**" **is** the **power** the energy that drives the universe, the ability and use of power of the gods, it is wielding of energy to bring about change, the power to animate substance, the ability to regulate the metering out of power.

Impression - a telling image impressed on the senses or mind. An Impression is a representation of the outstanding, prominent or striking features of a thing or object by using an artistic medium such as painting, writing, recording or a video (movie).

Individuation - In psychology it is the balancing of the opposite selves. One self is that we believe we are, the other self that which we believe we are not (The Shadow Self). These can be understood as the twin selves of a person. An example is: some men believe they must act manly, be macho, tell women what they must do, act rough and tough, and that women need be feminine, Nancy fancy. Psychologists have found both men and women are capable of either behavior, and both just need to carry out the type behavior as befits the situation. The result will be a balanced character that can be gentle when needed and aggressive or demanding when needed. This balancing is termed individuation and is said to be a life long effort, <u>Not allowing this balancing of behavior</u> between the two selves of a person has resulted in many unfair and selfish acts in the past, along with hurtful relationships, and emotional pain that results from such un-balance. Where there is true "Love", an effort is made to achieve balance, to achieve individuation.

Inertia - the laziness of a body. Inertia is resistance of a body to a change in it's direction of motion, or it's rate of motion. Inertia in this theory exists spherical standing waves, in other words in matter.
In this theory "Velocity" causes matter to continue in motion even after the force causing it is removed, and "Acceleration" gives matter the ability to rebound after being deformed. Inertia's magnitude is often taken due to the amount of matter present.

Innate Force - Newton proposed "The Innate Force" as the cause of "Inertia", but did not specify it's source. In this theory, in matter waves it is due to acceleration.

Interference - When electromagnetic Waves passing through one another in the vacuum of space have no effect on one another, they emerge from this interference zone unscathed. When matter, is present in the wave "interference zone", electromagnetic waves, exhibit "additive interference" or "subtractive interference" in the matter.

Ishvara – Is the internal ruler of all things including, ruler within the Universe itself, ruler within "The Self", and ruler within "The Mind". All gods are aspects of Ishvara.

Ishvara is popularly seen as a combination of the three gods of the Hindu trimurti (Brahma, Vishnu, Shiva). In this theory "Ishvara" occurs as a condition of Brahman. Brahman is potentiality and actuality, the unlimited and the limited. Brahman is eternal. Brahman is all.

Ishvara refers to things that are actualized a moment at a time, things that are born and because of it have a life span, they have limitation of actualized existence and limitation of form. The Svetas Upanishad says "Ishvara shines like the sun irradiating all." The Brihad Upanishad states Ishvara dwells in all living things and they do not realize it. This is so when Ishvara is seen as the power of limitation and therefore things with limitation have it.

Jillion – A Jillion is an indeterminately large number. The term "A Jillion" has been used in the sense of "Very, vary many".

Kama - one of the "worlds" of Aryan-Hindu metaphysics. Kama is the world of Desire, desire to have one's wishes and urges fulfilled. In this theory Kama is the "world" of on-going action, it is where Initiation of directed activity takes place, where the connectedness or soul of the the message of directed power is carried into substance, where power becomes force, force to mold the shape and character of substance, force to mold behavior.

Kinematic – Kinematics is a branch of dynamics that deals with motion as velocity and acceleration apart from considerations of mass and force.

Kinetics -the mechanism by which change is effected, chemical and physical change. The ultimate mechanism of change is the cycle.

Life – Life in this theory **is the activity of the soul**, (the activity of the connectedness). Life can be seen as having two halves. "**The life of the Higher Self**" a life of satisfaction, a life of ideals for motivations, lived in harmony with that which is "other", with that which is "not self". The other half is "**The life of the Lower Self**", the physical life, A life constantly incomplete, not fully satisfied, a life of need and desire, of physical mastery, survival, and reproduction, a life that knows pain and pleasure, hunger, greed, jealousy, and desperation. Each person needs to balance these two living selves, balance ideals with what is reasonable to do. This balancing is sometimes termed "being practical". The two lives are of equal importance, one does not exist without the other.

Love - In this theory "Love" has three forms. **Love as "the need to become whole"** that is the attraction of complementary opposites, to as it were bond together. **Love as "The Need to be Individual"** to be separate, the repulsion of likes, the appreciation of existence, and the urge to survive. Thirdly, **Love as a "need to share existence"** with one's complementary opposite self, in this theory this is achieved by alternation and repetition, each opposite taking it's turn at controlling existence, at being dominant.

Mass – A lump, a bulk of matter cohering together. Mass has inertia, weight in a gravitational field, and three dimensional volume. **In this theory "mass" is** a spherical standing wave structure **termed "Matter".**

Matter - Matter is a product of the creation process, commonly termed "physical substance". Matter is physical order as opposed to chaos. Matter has mass. In this theory matter is a spherical standing wave phenomena in primordial substance animated by energy. Energy molds primordial substance into spherical standing waves. Matter has properties such as "bulk" that is it occupies volume, it generates the gravitational force, has inertia and elasticity. Matter gives shape to our world.

Memory - Memory is the storing and sorting of sensory inputs, and experiences, recorded perceived facts, ideals, and beliefs. memory when accessed by the mind lets the imagination have a palette to paint with, to imagine what may come to be in future or what may have existed in the past.

Mind – (Manas) is the activity of the consciousness, the activity of the soul of the universe. Mind is not the soul, it is the activity of soul.
Mind is an action-reaction activity which evolved into an ability to reason abstractly, to be able to see itself, to understand, to infer, to intuit, to compare, to judge, to evaluate priorities. Mind takes physical sensory data and organizes the data in an effort to survive better than chance would allow. The activity of mind has been described as the activity of "Will, Wisdom, Action". I see the mind as a computing machine that assimilates, stores, and processes input as cycles of thought. Mind is both a physical material body and a process. **Emanuel Swedenborg**, (1688-1772, Swedish scientist, philosopher, theologian) termed Mind, "The World of the Rational" The Aryan-Hindus termed mind "Manas".

Momentum -The quantity of motion.
momentum = mass times velocity.

Myth – Is a story to teach or explain or promote the legitimacy of something. Word images are often used, which to be understood as originally intended need interpretation, for example the phrase "that's a horse of a different color" does not necessarily mean, the color is different but is used to convey the idea of "being different" in some way. "To rise from the dead" may mean to rise from spiritual death, to live more in the spirit, not just for the physical self.

Nodes - In a standing wave a node is a place of zero (or very near zero) undulation. There are both "Displacement Nodes" and "Velocity Nodes" in a standing wave. These are one fourth wavelength apart. A velocity node is simultaneously also a displacement anti-node. A velocity anti-node is simultaneously a displacement node.

Oracle – A means, place, or instrument by which hidden knowledge is revealed, or purpose made known. The mind is the oracle the inner seer and speaker to "The Self".

Oscillation - A moving backward and forward past a point of equilibrium, as does a pendulum or the tines of a tuning fork. The swinging from one extreme to another is Simple Harmonic Motion, an oscillation in which fluctuation of amplitude has a constant maximum, and displacement is proportional to, and oppositely directed to amplitude of acceleration.

Passive Energy is Tamas Energy, Tamas being a quality of passiveness, A thing with Tamas quality "reacts" only when acted upon. Passive energy is potential energy, is stored energy.

Permittivity - the toleration, or allowance of something. Electrical permittivity would be the ability to permit an electrical current to flow.

Phase - the fraction of a period that has elapsed. Example: In a Simple Harmonic Motion wave acceleration is zero when displacement is zero, but velocity does not reach zero until a quarter wavelength later. Velocity is said to be $1/4^{th}$ of a wavelength out of phase, and since it reached zero later, after zero displacement, velocity is said "to lag" displacement.

Potential Energy – is stored energy. Potential energy is the complementary opposite of Kinetic Energy. Potential energy is passive energy. Examples of potential energy:
An object raised in a gravitational field has potential energy.
Energy in a compressed steel spring is potential energy.
In this theory the elastic energy, the strain in substance due to it's displacement is potential energy. The adhesive "gluonic wave field" that holds an atomic nucleus together contains potential energy.

Power – The use of energy over a duration of time. Power is the rate of doing work.

Power = Time x force x distance

Practical – The state of actually being engaged in an activity or attempting to accomplish something.

Pressure - Pressure is a force applied over an area. English physicist John Henry Poynting (1852-1914) found the pressure of sound waves is due to their momentum. Momentum being the mass of the particles times their velocity.

Primordial Connectedness is a Carrier, Energy into Substance.
In substance the message becomes a force.
In the matrix diagram of the physical universe, "the message" is one of "Velocity" and "Acceleration", these animate and shape Primordial Substance. **Other Names for the "Connectedness" are "Primordial Soul" or "Primordial Consciousness".**
"a Messenger" that delivers

Pseudo-Substance – that which resembles a substance however it is not quite a substance, but more like substance than a property. Newton's space has been likened to a pseudo-substance, In this theory Space and Time are the characteristics of primordial substance, they are both substances.

Psychology the science of behavior, the study of the interaction between a biological organism and it's environment, also the interaction between individuals or an individual and society.

Psyche - The personality. Psyche is the psychosomatic unity of mind-body, the vital principle, of a substance, the "livingness" of a substance. A person's "psyche" is "the self. The Psyche is the, cognitive, and conative (that is the biologically instinctive reactions to conditions that bring on an urge, desire or reflex action for change), and the emotional aspects of a psychosomatic unity.

Psychosomatic - the interaction between mind or emotion, and body.

Purusha – The Man, The Self, The Person, the view taken is "a man is what he does and says", in other words a man is his behavior. In the theory in this book, Purusha is the message delivered by the soul.

Rajas - Is one of the "Gunas", the "Qualities" of Aryan-Hindu mythology. Rajas is aggressive impatient behavior. Rajas Energy is the energy that initiates change (lower right corner square of physical universe diagram, pages 220-221). Rajas Energy is the Energy of animation, the Kinetic Energy.

Rarefiable – capable of being rarefied. To rarefy something is to make it less dense, Porous, or more scarce. To rarefy a gas is to lower it's pressure.

Reason - the power of comprehending, a thinking process whereby a goal is reached. "Reason" makes sense of diverse facts or sensations in a most basic way. In the theory in this book one of the main operations reason does is sorting out all the sensations, and experiences the body receives to find pairs of complementary opposites, these are the extremes of their kind, such as light-dark, substance-non substance, permanence-change, yes-no, same-other. The extremes are then used to make value judgments of in-between conditions.

Reflection - The action of bending back, the partial or complete return of wave motion on encountering a reflective surface, a different medium, or a different optical density. A reflector curved like a bowl with a parabolic curve will focus parallel waves striking it. This is how a reflecting telescope gathers light.

Refraction - The bending of waves as they pass from one medium to another, as when they pass from air into water, or glass into air. This bending of the waves as they enter a glass prism, slows their speed and deflects them through an angle, the greater their frequency, the greater their deflection as they emerge from the other side of the glass into the air, where this refraction phenomena lets white light be separated into the colors, to form a band, a series of colors. Red waves are deflected least, then Orange, yellow, green, blue, purple, and violet. Violet of these is deflected the most. This series forms a continuous spectrum band of colors.

Reincarnate – to make anew or repeat again, to be reborn. In myth this rebirth represents putting illumination in place of ignorance, living understanding in place of dead facts and ideas.

Reluctance - "reluctance" is an unwillingness to permit change, a resisting of change, such as opposition to certain behaviors not in accord with those of society or one's personal behavioral choices. In electronics reluctance is a delay in the build up of a magnetic field to full strength.

Representation Is the showing, depicting, or portraying of a likeness of a thing in such a way that the mind forms an image or idea of the object.

Resonance - To resonate is to echo, or resound, to repeat. Periodic motion or displacement of waves, as an echo, a reflecting back and forth on itself, is a resonance. Under certain conditions repeated resonant oscillation occurs of electric and magnetic fields in electromagnetic waves. There is a resonance in standing waves due to reflection between two limiting zones, the distance between these reflection zones is critical to permitting resonant behavior at a given frequency.

Sattva -Sattva is the "quality" termed harmony. Sattva is a sharing of existence in the universe. Sharing of existence is accomplished by alternation and repetition. In psychology Sattva, is the harmony of the two selves, each allows the other to dominate behavior as needed.

The Self – Is the awakened sense of, "I Am", "I Exist". All outside me is "Other". Self image is one's own opinion of one's behavior, appearance, or importance.

The Shadow Self – Is the twin self of psychology. The Shadow Self is the self one believes they are not. Certain behaviors are thought to dominate in males others to dominate in females, as their primary behavior, aggressiveness in males, gentleness in females. However psychologists have found both behaviors exist in either gender (sex). The behavior that dominates is a matter of the situation present, and the presence of chemical hormones, these cause bodily urges, and can pressure a person to satisfy the urge. A balancing of these two selves is needed to produce a reasonable, amiable personality. The balancing process is termed "individuation".

The Soul is also termed **Prajna.** - The soul of the universe is "Consciousness", is "Connectedness". The Soul is the messenger that carries the message of energy into substance. In physics the message of the "soul" would be in the form of "Velocity" and "Acceleration". In an animal the message of energy carried by the soul is the power of animation, the delivery of the forces of "life". Since substance and energy are eternal "the connectedness" between them, their "soul", is also eternal. It is just a matter of what message the soul is carrying, animation of body, or animation of mind.

Space a volume with three dimensions of extension. In the theory in this book "Space" is one half of Primordial Substance, the other half is "Time". Space is a container of things yes but it also has attributes when displaced. Space gives "things" a place "to be". Space allows things to have "Form", and fields of force. When space is displaced "strain forces", elastic forces occur within it. Both space and time in this theory are "Substance".

Spirit (Primordial) - is **Prana**. Spirit is Energy, Spirit is the power that animates substance, that powers it's livingness, it's behavior. Spirit is, not breathing, it is the power that drives breath. The spirit of the mind is "The Will". The spirit of substance is energy.

Speed - rate of motion without specifying any direction.

Sthula – Animated physical substance. The body is animated by energy, the animated substance then is said to have behavior. Sthula in the theory in this book is animated space substance and animated time substance, their animation takes the form of changing amounts of displacement.

Strain forces in this theory these are forces due to "Acceleration" acting in primordial substance to resist displacement that is increasing and aide displacement that is decreasing. The "Strain forces" in substance are the result of displacement.
The standard physics text definition of strain is: the deformation of a body divided into it's original dimension before the body was deformed.

Stress forces in this theory "Stress Forces" are due to "Velocity" acting on primordial substance. Stress forces drive substance into displacement. Stress Forces mold substance that is form substance, like a potter molding damp clay. "Velocity" in this way gives form to substance. The physics text definition of stress is: a Force per unit area that tends to produce deformation.

The Subconscious Is the part of consciousness that is a background against which thought takes place. The subconscious is memory, reason, and understanding, especially understanding of "things" as word definitions, these are the ideals of one's existence, from which are selected goals to accomplish. **It is in the sub-conscious that one believes what one "believes".** The subconscious is where our personal inhibitions and those of society are known. The Subconscious is the place of dreams and the place hypnotists wish to reach with their suggestions, where they tend to register as fact.

Primordial Substance is **Prakriti.** Prakriti is eternal, indestructible homogeneous substance that pervades the entire universe without any voids. In this theory "Space" and "Time" are both substances and can be displaced. Displaced Space becomes the forms and force fields of our world. Displaced time becomes actualized "Duration".
"Primordial Substance", is one of three basic components of the universe, the other two are "Primordial Energy" and "Primordial Connectedness". Substance is not Matter until it is "formed" into matter by energy, just as moist clay is a lump and not a vase until it is formed.
See page #84, The Nature of Substance.

Tension - tautness, the act or action of stretching or the condition or degree of being stretched. The opposite of tension is relaxation. Tension is the elongation of an elastic body. Tension is achieved by displacement of substance, in a Torsion Wave, tension is achieved by a twisting motion. In the world of art tension is a balance maintained between opposing forces, such as light and dark, opposing colors, or between square shapes and round or curved shapes.

Time – Is unlimited duration, in a word "eternity". Time when limited is "life-span". In the theory in this book, Time is one of the two parts of Primordial Substance both are real. Un-displaced time is the existence of potentiality. Displaced time is on-going actuality. one half of time is un-moveable, fixed, the other half of time is movable, displaceable compared to it's un-moved counter-part's position. When displaced, time acquires a strain, this "strain" has the power to return time to it's un-displaced state. See page #95, The Nature of Time.

Understanding is the end product of reason. It is the answer. On the matrix diagrams understanding is the reasoned-out word definition of a thing, this definition is seen as the "Ideal" of a thing, the next best thing to the actual thing itself.

Universe - Universe in this theory is the totality of all that exists, has existed or will come to exist. Universe is eternal and cannot be destroyed. There is only one Universe existent. There may be a succession of alternate, universe conditions, each however is due to the changing of a single whole. All the parts of the universe are connected to all other parts, by "The Connectedness", which is also termed "The Consciousness", or "The Soul". Universe has two extreme states, Chaos (disorder) and Cosmos (order), and all proportions in-between of these extremes come to exist over time. Extreme Chaos can be viewed as extreme diversity. Extreme order (cosmos) can be viewed as a singularity, a unity.

Velocity - the rate of change, the amount of change per unit time, the rate of "flow", or rate of displacement. On a graph of the three harmonic wave motion characteristics, velocity is one fourth wavelength out of phase compared to displacement. Velocity reaches maximum in the wave time wise before displacement and is said to "lead" displacement maximum. Velocity and Displacement have an inverse relationship in that, as displacement increases, velocity decreases. Velocity makes possible the overshooting of the mark by the wave motion of substance.

Virgin Birth – the birth of spiritual life in the heart.

Vis Viva – Is the living force. In 1807 **Thomas Young** (1773-1829 English physicist and physician) changed the name from "living Force" to "Energy".

The Waking Consciousness – Is our awareness, of ourselves, one's situation, and surroundings when in the waking state. Waking Consciousness is the state when not asleep.

Waves – Undulation of magnitude of displacement, or elastic deformation traveling through substance. A wave is an oscillatory motion of substance. If a wave could speak it may well say "My name is legion for we are many" Mark ch.S, vs.9.

Wisdom - Wisdom is balance. In the world of psychology it is a balance of the self and the non-self (the shadow self). In the world of thought it is an ordering of priorities and letting the most important be acted on first. It has been said that "Wisdom" is more precious than gold (material wealth).

The World Center – The Axis Mundi. That about which the world as it were revolves. In Meta-physics the center of the world is "The Self", the place where the higher and lower selves meet. In the world of physics it is where the potentiality of "form" and the actuality of "The Now" meet.

The World Soul – <u>Here</u> it is understood as **Prajna**, the Soul of the Universe. The Soul of each thing in the universe is a part of a whole, "The World Soul". The World Soul is also the "Primordial Consciousness" of psychology, and it is the "Primordial Connectedness" of the physical theory in this book. The World Soul is eternal and is one of the three basic components of the universe. "**The navel of the world**" represents that from which life is sustained, the spiritual life, and the physical life, the life <u>is the activity of the soul</u>.

Glossary of Egyptian goddesses and gods
See page 121 for their matrix diagram

Anket The Goddess Anket, is the complementary opposite of Khuum in that she is the Goddess of the lands of the two energies, the potential and the actual (Kinetic). Anket balances these energies in the universe by alternating the prominence of one then the other, repeating this process over and over, in this way both energies co-exist as equals. When conditions allow, the two energies will oscillate, will resonate. In this way the energy is controlled. In the world of psychology, Anket is the balancing of motivations (directedness) of the "Primordial Will" of "the Self". When this results in a balance both the selves co-exist in harmony, exist in a condition of Sattva.

Atum The God that made heaven and Earth come to be. Let us say Atum is "Chance", is "Probability". Chance and probability allowed the conditions within chaos to give "Substance" more distinct forms.

The Ba – Prajna, is The Primordial Soul, The messenger of the Gods. The Ba Carries the power of directed energy into substance to form substance. The message carried by the directed forces, do the forming and allow the dissolution of form. The message carried by the Ba brings substance into activity, bring substance to life.

Enki the God of Opportunity.
Chronos, Father Time Sin.

Isis (Auset) The Queen of Heaven. In this theory Isis is "Space". Isis can inhibit or encourage change just as acceleration does.

The Guardian of Destiny – Enlil, Shamash "The Judge of the World", fatherhood.

The Goddess that gives favor - Each culture's name for her may differ, but she says who sexual favor will be granted to. To the Sumerians her name was **Innana**, to the Assyrians and Babylonians she was **Ishtar**, to the Egyptians she was **Isis**, and as **Astarte** she was a goddess of the Semitic pantheon. King Solomon built for her in a high place (1 kings ch 11,vs. 5-7 and II Kings ch.23, vs.13). The hard liners did not like this, it could mean some loss of their power over the people.

The Ka - The Ka in this theory is the message carried by the Ba. The Ka is the directed forces.

Khuum - The God of the two lands. The two lands here spoken of are the land of permanence and the land of change, (impermanence). These are the "worlds" of Atma and Kama in the Aryan-Hindu scheme of things. Khuum is pictured in Egyptian art as standing by two identical human figures, twins, the one is the impermanent, the other the permanent (the "Ideal"). When one is "molded" so is the other for the "Ideal" is the actuality of the potentiality of a thing. Ideal here does not imply the thing is bad or good, only it is possible. See diagram page #121, Fig. #16, in the left vertical column center square the two lands co-exist (this is square with Khuum's name). Khuum is in the Aryan-Hindu world of Sthula, as are Isis and Osiris.

Lachesis, Atropose, and Clotho, These are the three "Fates" of ancient Greece. Usually depicted as three Women, between them they gave each person a measure of good and evil, a life span, and an inescapable Fate.

The Goddess that brings unity the Mother goddess **Lamassu** "The Womb that gives form", and **Ninhursag** "Lady Birth", motherhood.

Nephthys - In Egyptian mythology she is the daughter of Nut, and sister of Isis. She is "The Lady of the House", and as "Mut" motherhood. In this theory Nephthys is the attraction of opposites to produce a whole. In the mind she is "Reason".

Nu (Nun) The Cosmic Ocean. (Brahman), Universe.

Nut Goddess of the sky. Nut the ability to control, to awaken. In this theory Nut is "Primordial Energy".

Osiris The God that is Time in this theory, eternal existence. In the Egyptian myth Osiris is murdered, and resurrected as **"Horus"** his son.

Ptah Is the God which is the Potter that molds the land of impermanence, the land of on-going change. Ptah (Phtas) - The God of the Forces that mold substance, the forces of directed change. The Ptah is the complementary opposite of Sekmet.

Ra (or Re) The hidden force of the cosmos. Ra is the Sun God, the God who dies in the west at sunset and is carried across the night sky to be reborn at sunrise each day. Egyptians of the time of the pharaohs believed the goddesses and gods guided and directed the natural forces of the universe. **Ra** in this theory **is the activity of the soul, the Ba.**

Seb - (or Geb) The God Seb, is God of Earth.. In this theory Seb is the underlying substratum of all things. Seb is "Primordial Substance", the "something" that when acted upon by energy is "formed" into the things of our world. Here Seb is the Prakriti of the Aryan-Hindus.

Set (Sutekh) - In this theory Set is the ability to direct the power of the universe, to instigate change.

Sekmet – "**The Great Cat**", the Egyptian equivalent of **the lionesses** of the Sumerian Goddess "Inanna" walked with the lioness on a leash. Sekmet in this theory is the forces that inhibit or encourage change, much as acceleration does in the world of physics.

Shu - Shu is the complementary opposite of the Goddess Tefnut. Where Shu is the dryness of the desert, Tefnut is the moisture of the fertile plain.
Shu in this theory is the world of change (the Aryan-Hindu equivalent world is Kama). Shu is a world of Actuality, of instigated on-going directed forces of change. In Shu there is always incompleteness, on-going need and desire.

Tefnut - Goddess of the Divine World of permanence. Tefnut is the heavenly world of satisfaction, of "Ideals", of "Permanence". Tefnut is a world of potentiality. The Egyptian Tefnut is the "Atma" of the Aryan-Hindus.
Tefnut and Shu are complementary opposites.

Four of these Goddesses and Gods (two each),
I believe are the four phases of a cycle of change. The four
phases of this repeating cycle are:

Tefnut-(Atma), Nut-(Buddhi),
Shu-(Kama), and Seb-(Sthula).

The colored pigments Egyptians used in their paintings at
the time of the Pharaohs, there were ten of these:

White Lead, Smoke Black, Malachite green
burnt umber, Egyptian Blue, Red Earth,
Brown Earth, Violet Earth, Grey, and
Dark Yellow.

Notes

An after word on the dream of Universe

The sages recognized that duality must be part of the universe. Duality is pairs of complementary opposites, and a form of static balance and of symmetry. All theories of the universe therefore need duality as a basic part of their design.

In the myth of **the dream of universe, Vishnu** sleeps in the Cosmic Ocean, **his dream is an ability** to balance the forces of the universe, Vishnu is the sustainer god is the ability to balance the power of creation and the power of dissolution. To balance the wills of Brahma (creator) and Siva (destroyer).

Brahma sits on a lotus blossom and Wills order in the universe. **Shiva** from within the Cosmic Ocean wills dissolution. I have included Shiva even though no mention is made of Shiva in the myth. **See Fig. #46, page 267.** Now the two complementary opposites Brahma and Shiva can work in harmony thanks to Vishnu. Brahma gives birth to order "Cosmos" and Shiva undoes the work of Brahma driving cosmos into dissolution, back to chaos. Shiva removes the old and makes way for the new. Brahma supplies the new. See also pages 9 and 10. The three gods are personifications of the powers of nature, they are powers of energy control. The three gods are existent characteristics of the cosmic ocean, the energy of their "Wills" molds substance.

If I were to place the three gods in a matrix diagram, all three would be in the right hand Buddhi column, with Brahma in the top square, Brahma gives birth to the new and mothers the new. Vishnu would be in the center square, the wisdom of balance, the wisdom of coexistence. I would put Shiva in the bottom square. Shiva destroys the old and in so doing activates opportunity for the new, like a fatherly "provider".

The dream is a vision of how a universe of order appears out of chaos and eventually sinks back into chaos all due to energy control.

The following illustration Fig. #46 may help picture the dream of universe in your mind.

The Myth
Brahma Creates The Universe of Order
here Shiva is also present

Fig. #46

Index of Persons
mentioned in "I Am Universe"

Anaximander (Milesian philosopher)   81
Appolonius (from Tyana)   70
Aristotle (Greek philosopher) 70, 95, 98
Aspden, Harold (British electro-physicist,
                        inventor)   72

Bain, A (Scottish philosopher)   125
Bartocci (Italian professor)   115
Bogart, Humphrey (U.S. Actor)   108

Cather, W.S. (U.S.A. Novelist)   18
Cerenkov (Soviet physicist)   59
Cicero (Roman writer scholar)   103, 241

Dobado, A.   90, 151
Dolbear, A.E. (Physics and astronomy
                professor)   73, 125
Descrates (French philosopher)   71
Doppler, Christian (Austrian physicist) 103
Donne, John   105

Einstein, Albert (German-Swiss-U.S.
      mathematical Theorist, Physicist) 72, 113, 115
Empedocles (Greek philosopher)   124
Ezekel   12

Gray, Hobart (U.S. Major General)   174

Halley, Edmund (English astronomer) 112
Hatch, Ronald (U.S.A. Global positioning
                    scientist)   72

Index of Persons continued

Heraclitus (Greek philosopher)   16, 63
Herbert, G (English poet)   109
Hesoid (Greek poet)   46

Job   12
John   16 17
Joubert, Joseph (French philosopher) 84

Kelvin   71
Kepler, Johannes   71
Khendjer (Egyptian pharaoh) 97

LaFreniere, Gabriel (Canadian physical
              theorist)   72,74, 103, 157
Lagrange, Joseph Louis   53
LaRochefoucauld   234-235
Lodge, Sir Oliver   71, 126
Lorentz, Hendrik (Dutch physicist) 72
Loser (Egyptian pharaoh)   28
Lucretius (Latin philosopher)   91

Maroto,A.L.   90, 151
Maxwell, James Clerk   71
Mendeleyev (also as Mendeleef, Russian
              Chemist)   158)
Muhyedeen, Bhajat Jaafar   113-115

Newton, Isaac   7, 71, 112
Nodland, Borge   90, 150-151

Opik, Ernst J, (Estonian Astronomer) 192
Ovid (Roman poet)   46, 95

Index of Persons continued

Paramenides (of Elea Grecce) 62
Patton, George Smith (U.S. Army General) 173
Pepi I, 28
Plato (Greek philosopher) 95
Poincare   7, 8
Poynting   251
Pretto, Olinto de (Italian industrialist) 115
Publilius, Syrus (Latin Writer) 109

Reich, Wilhelm (Austrian scientist) 72
Ralston, J.P.   90
Ramses II (Egyptian pharaoh),   28
Russel, Walter (American Polymath)   75

Shakespeare   8, 108
Sharrum-kin (Sargon) of Akkad   29
Soddy, Frederick (English radio-chemist) 125
Swedenborg, Emanuel   249

Tennyson (English Poet)   96
Tesla, Nikola (U.S. inventor)   7, 72
Thomson, J. J.   165

Vind, Charles (fluid space)   75
Virgil (Roman poet)   98, 109

Watt, James (Scottish engineer)   110
Wolff, Milo (U. S. Mathematical
                              physicist) 75

Young, Thomas (English physicist) 71

Index

Acceleration 122, 206, 239
Active energy Kinetic energy 236
Aether 69
Aker Symbol (Egyptian) 93-94
Alter Ego 239
Anti-Gravity 175, 177, 179
Aryan-Hindu "Creation Hymn" comments 31
Astronomical Distances 201
Atma 37, 39, 240
Atom, size-165, half-life-160, isotope-161
        atomic mass-161, atomic number-159

Ba and Ka 117, 119, 260, 261
Balance and the Universe 54
    Static 54, Dynamic 54
Imbalance, controlled 57
Balance as Harmonic Motion (Graph) 60
Basic Components of the Universe 66
Belief - 240
Body 240
Brahma 10, 11, 234, 266
Brahman - 240
Buddhi 240

Cerenkov Radiation 58-59
Chaos and Cosmos 192
Compassion 241
Complementary Opposite Pairs 63-64, 127
Concluding comment 209
Consciousness (Primordial), is Prajna, 83, 118, 241
    Subconscious 83 118, 257
    Waking Conscious 259

Index continued

The Id   83, 119, 245
Cosmic Ocean 9, 20, 21, 22, 24, 241, 266
Counter Earth   241
The Creative Power and Gender   234
Cycle   46, 192, 241
   Cycle of Change, 4, parts of 36-39, 44, 48
   Phases of Cycle   189-190
   Cycle of orbiting Body,
        earth-sun   202, moon-earth   203
   Harmonic wave graph of an orbit? 205-208
Cycle of Social Change for Man   46
The Counter Earth   53, 241

Deity   242
Description   241
Desire -   242
Diffraction -   242
Displacement 52, 106, 112, 128, 206, 242
   of standing waves   141, 144
Duration -   243
Dynamis and Energia   111

The Ego   243
Elasticity   243
Electromagnetic Wave   35, 132-133
Electron (Natural)   57, 167
   Positron   157
   Artificial Electron   166
Elements (The 5 Ancient Elements)   68
Elements (Modern Chemical Elements) 158
   Atomic number   159
   Atomic Mass   161

Index continued

Isotopes   160
Periodic table of chemical elements 158
Elements dissolved in sea water   163
Energy (the nature of)   108-115, 243
  Kinetic Energy (active energy)   111, 112, 243
  Potential Energy (Stored Energy) 111, 243, 250
  E equals m c squared equation   113
Equality   244
Either (Aether ancient element)   69, 75
Exponential Notation and the SI system 210-211

Firmament   30
Fields of Force (The Gluonic Field) 157-158
  Forces of stress and strain   123, 256
Force   244

Galaxy – Milky Way   193
Glossary of Terms (Definitions)   239
    Goddesses and Gods of Egypt   260
    Goddesses and Gods of Sumer   26
Gravity   169, 179
    Cause of Gravity   175
    Gravity Well 184, 187
    Gravity Force, on Various Planets 180-184
    Gravity and Space Ships   92, 172-174, 177
Gunas (The 3 Qualities)   49
    Tamas 49, Sattva 50, Rajas   49
    Gunas and Simple Harmonic Motion (Graph) 52

History   244

Id   245

Index continued

Ideal Forms   245
Image   245
Individuality (metaphysical Higher Self)   232-233
        Personality (metaphysical Lower Self) 232-233
Individuation   245-246
Inertia   112, 155, 246
Innate Force (Newton's) 73, 112, 155, 246
Inverse Square Law   184
Ishvara   215, 246-247

Ka and Ba   117, 120, 260, 261
Kabala (Hebrew)   11
        and Balance   56
Kama   247

Lagrange equilibrium points   53
Life   235, 248
Logos   11, 12, 16
Love   235, 248

Map, Ancient Sumer 19
Mass   157, 248
Mathematical handling of scalar
        and vector quantities 152
Matrix Diagrams
        On the Construction of 214-218
        Goddesses and Gods of ancient Egypt 121
        Physical Universe   220-221
                Size of   91
        Metaphysical Reality   222-223
        Mythological World   224-225
        Mind (Manas)   226-227

The Psychological Self 228-229
    of the Sumerian Goddesses and Gods   24-25
Matter   92, 157, 162-163, 244
Mazzaroth   13
Milky Way   193-200
Mind   245
Myth – Brahma Creates the Universe 9, 265-266

Oracle -   9, 245

Persona   229
Phase   189, 250
Pillars (or supports) of the universe 36-45
Primordial Connectedness   251
Primordial Substance   252
Purusha   83, 252

Rajas   252

Sattva   254
The Self (Actual as the Id)   229, 254
    Shadow Self (subconscious as a reaction) 255, 252,
    as The Double   231, 255
The Self, Higher and Lowe   46, 248
Simple Harmonic Motion (Graph)   52
Shiva   11, 266
Solar System   180-181
Soul   116, 255
        Divine Soul, Human Soul, Animal Soul 118-119
Space (the nature of)   86, 255
Spirit   110, 256
Sthula   256

Index continued

Standing Waves   137
     Standing Sound Wave (longitudinal
          Wave)   138, 154
     Spherical Standing Wave like that
          of the electron   146-147
     Spherical Standing Electron Wave 146, 147
     Anti-Gravity Producing Spherical
          Standing Wave   179
Subconscious   257
Sumerian Concept of Creation   23
Sumerian Concept of Universe   20
Sumerian Goddesses and Gods 21-28
Sumer (Map)   19
Substance, displacement of
   (in this theory) 106, 112, 128, 257

The Tao   127
Tension   257
Time (the nature of)   95, 258
     Absolute Time (the time of Newton) 102
     Time Continuum   95
     The Now of time 95, 98
     Other classifications of Time 102, 104
     Relative Time (Doppler Effect)   103
The Twin Selves (The Double etc.)   230

Universe   258
     Three Components of Universe   66
          Prakriti, (substance) 80-83
          Prana (Spirit),(Will) 80-83, 110
          Connectedness, Soul, (Prajna) 80-83
          Energy, (Prana)   80-83

Index continued

Universe (continued)
    Substance, (Prakriti) 80, 83
    Will   109
    Correspondences of terms for (table) 82
    Universe of Complementary Opposites 64
    Boundary of Universe   91, 103
    Name of   78
    Oscillating Universe   192
    Universe Wave   90
    Substance of (Prakriti) 80, 85
    Supports of the Universe 37, 48
    Tension in the Universe 63, 145
    Mostly Un-provable Characteristics of the universe 61
    The Tangible Universe   68

Velocity   112, 206, 258
    Escape Velocity 182-183
Virgin Birth   259
Vishnu   9, 11, 266

Waves in the Cosmos 127
    The three characteristics of waves 130
    Location of wave characteristics
    in a matrix diagram or graph 131
    Electromagnetic waves 35, 132-133
    Handling of waves   154
    Longitudinal waves   138, 154
    Speed of waves   135-136
    Spherical standing waves 147
    Standing Waves   137
    Wave nodes and anti-nodes 137, 142, 239, 250
    Torsion waves   150, 151, 153

Index continued

Waves in the Cosmos (continued)
   Transverse waves   152, 154
   Traveling Waves   127, 130, 136
Wave Phase   189
Will   109
Wisdom   223, 259
World Soul   259
World Systems or Planes 34, 36, 45, 214
   Paran-atma   34
   Atma   39, 240
   Buddhi   40, 240
   Kama (The Astral world) 41, 247
   Sthula   42, 256
   Manas (Mind) 44, 226-227, 249

Yin and Yang (Chinese symbol) 65, 127

**Location of subjects mentioned on back
cover of "I Am Universe".**

The working electron. 165-167

The wealth held in sea water.   163

The possibility of the boundary of the universe acting like a mirror. 91-92,

The universe seen as the dream of a god. 263

Is Your Shadow calling you? 231

The World where all things are ideal, Atma 39

La Rochefoucauld tells why there is love 235

<u>Ordering Information</u>

Additional copies of "I Am Universe" can be ordered in the U.S.A or Canada in one of four ways:

1) By purchasing from your local book store.

2) By calling Infinity Publishing at:
       1-(877) BUY-BOOK
       1-(877)-289-2665

3) By visiting the web site
   www.InfinityPublishing.com
       or at
   www.BuyBooksOnTheWeb.com

4) or from – www.Amazon.com

Note: Order from Infinity Publishing and
      get 40 percent off on orders of five
      books or more.

Made in the USA
Monee, IL
07 July 2026